Web 前端技术丛书

Angular 开发实战

李一鸣 著

清華大學出版社

北京

内 容 简 介

前端开发市场火热，Angular 称为前端三大必学框架之一。本书是一本为学习 Angular7 的开发人员量身定制的快速入门教材，书中实践案例多，适合喜欢边学习边动手实践的读者。

本书分为 12 章，内容包括 Angular7 开发环境、TypeScript 语法、指令、组件、依赖注入、HTTP、表单验证、路由、自动化测试、json-server 后台模拟环境的搭建、待办列表实战、商城后台管理系统实战等。

本书内容详尽、示例丰富，适合 Angular 初学者，同时也适合高等院校与培训学校计算机相关专业作为教材使用。

图书在版编目（CIP）数据

Angular 开发实战/李一鸣著.—北京：清华大学出版社，2019
（Web 前端技术丛书）
ISBN 978-7-302-53482-2

Ⅰ.①A…　Ⅱ.①李…　Ⅲ.①超文本标记语言－程序设计　Ⅳ.①TP312

中国版本图书馆 CIP 数据核字（2019）第 172245 号

责任编辑： 夏毓彦
封面设计： 王　翔
责任校对： 闫秀华
责任印制： 李红英

出版发行： 清华大学出版社
网　　址：http://www.tup.com.cn，http://www.wqbook.com
地　　址：北京清华大学学研大厦 A 座　　邮　　编：100084
社 总 机：010-62770175　　邮　　购：010-62786544
投稿与读者服务：010-62776969，c-service@tup.tsinghua.edu.cn
质 量 反 馈：010-62772015，zhiliang@tup.tsinghua.edu.cn

印 装 者： 北京鑫丰华彩印有限公司
经　　销： 全国新华书店
开　　本： 190mm×260mm　　**印　　张：** 12.75　　**字　　数：** 326 千字
版　　次： 2019 年 10 月第 1 版　　**印　　次：** 2019 年 10 月第 1 次印刷
定　　价： 59.00 元

产品编号：080796-01

前　　言

读懂本书

未来前端开发有发展前途吗？

前端开发目前在国内非常火，随着前端开发市场持续扩大，不仅仅是网页，甚至移动端、PC端应用都可以使用前端技术进行开发。现在国内的一线大公司都已经使用前端技术来开发自己的手机 App 了，比如微信小程序、淘宝天猫、京东、饿了么等，其火爆程度一目了然。

——现在入门还来得及吗？种一棵树最好的时间是十年前，其次是现在！

你还在用 jQuery 吗？已经过时啦！

虽然有很多旧网站还在使用 jQuery，但是大多数公司的招聘标准早已改变。前端开发在这些年飞速发展，技术更新速度很快，如果想保持自己的竞争力，必须持续保持学习状态。现在想在前端开发领域找到一份好工作，则必须学好 Angular、React、Vue 三大框架！

——各大招聘网站都可以搜一搜看一看，要不要学，你看着办！

Angular 在前端开发中有哪些优势？

Angular 作为一个比较大而全的前端框架，它使用 TypeScript 进行开发，包含服务、模板、双向绑定、路由、依赖注入等各种便捷的功能，可以让你的开发效率事半功倍。Angular 由 Google 公司开发并维护，有着众多的开发者和活跃的社区支持，而且中文文档翻译十分全面。

——三大框架选择困难怎么办？小孩才做选择，成年人全部都要！

本书真的适合你吗？

只要你拥有一定的前端开发基础或者有 AngularJS 的开发经验，那么本书就能帮助你顺利入门 Angular7。从环境搭建到 TypeScript 语法讲解，从内置指令到组件表单的每一个知识点，我们都手把手地教你学会。

——知识点太多怕学不会？不用担心，每个章节都配有例子帮助你理解关键的知识点。

本书涉及的技术或框架

- HTML
- CSS
- Chrome 浏览器调试
- Angular
- HTTP

- Karma
- NPM
- Git
- JavaScript
- SCSS
- JSON
- Ng-Zorro
- HTTPS
- Protractor
- Node.js
- Visual Studio Code
- TypeScript
- LESS
- Json-serve
- Ng-Alain
- Jasmine
- Postman
- CNpm

本书涉及的示例和案例

- Hello Angular
- 自定义结构型指令
- 路由框架的搭建
- 单元测试常用 API
- 城市组件
- 用户信息页
- 端对端测试常用 API
- 使用 json-server 实现增删改查
- 待办列表
- 商城后台管理系统
- 制作一个 HTTP 拦截器
- 在登录组件实现模板驱动型表单
- 使用响应式表单构建个人资料页

本书特点

（1）实例为主、理论为辅，本书以理论知识的介绍为辅，以大量代码示例为主，通过众多精心选择的典型例子，帮助读者更好地理解 Angular7 开发中的重点、难点。

（2）循序渐进、轻松易学，本书的章节安排由浅入深，从简单的知识点开始，一点点增加难度，激发读者的阅读兴趣，让读者能够真正学习到 Angular 的使用技巧。

（3）技术新颖、与时俱进，采用目前最新的 Angular7 版本，避免学习旧版本导致知识不通用。结合时下热门的技术，如 Node.js、json-server、Ng-Zorro 等，让读者在学习 Angular 的同时，了解熟识更多相关的流行技术。对于无法全面讲解的一些框架，还给出了官方文档的详细网络地址以便于读者上网查阅。

（4）贴近读者、结合实际，书中使用的 UI（用户界面）框架、自动测试框架等，都是实际开发中常见的、使用率高的框架，保证读者可以学以致用。

本书读者

- Angular 开发初学者
- 前端开发工程师
- 前端架构师
- 从事后端开发且对前端开发技术有兴趣的人员
- 想学习 Angular 自己开发网页的人员
- 可作为高校和培训学校相关专业的 Web 前端开发实践教程

代码下载

本书示例代码下载地址：https://share.weiyun.com/5zjTnFL（注意区分数字与字母的大小写）。如果下载有问题，请联系 booksaga@163.com，邮件主题为“Angular 开发实战”。

作者

2019 年 7 月

目 录

第 1 章

初识 Angular

本章旨在为 Angular 初学者介绍该框架的基本发展情况，以及如何从零开始搭建开发环境，再到运行自己的第一个程序，让读者对 Angular 框架有一个初步的印象。在读者对 Angular 有了一定的了解后，最后笔者将会对目前前端各大框架的现状进行一些分析，并提出学习建议。

本章主要涉及的知识点有：

- Angular 简介
- 从零开始搭建 Angular 开发环境
- 实战练习：运行自己的第一个 Angular 程序
- Angular UI 库

1.1 Angular 简介

随着 Node.js 的诞生，以及 HTML 5 标准的确立，前端开发的热潮和各大网站的招聘使得前端开发人员供不应求。从当年的原生 JS（JavaScript，简称 JS）到 jQuery，再到现在的三大框架（Vue、React、Angular），互联网前端开发的门槛也在逐渐降低，越来越多人加入到前端开发者的行列。三大框架的流行，自然会有不少人愿意去主动学习，本书出版的目的就是帮助更多想学习 Angular 的前端开发者，通过更多的练习去掌握这门框架。

很多开发人员会自嘲为“码农”，觉得总是需要学习新框架、拼命干活十分枯燥。2018 年被大家称为“互联网寒冬期”，大量公司都有裁员的情况，很多大公司的开发岗位也逐渐减少。可能有很多程序员为市场上这种情况感到焦虑，这里笔者给出一个建议：对于我们开发者而言，既然无法改变大环境，那只能去适应它。在这种时期，我们需要做的还是不断提高自己的技术积累，技术够了就不怕没有合适的岗位。前端开发作为这两年比较火热的岗位，在各大招聘网站的开发岗位中，招聘数量还是比其他方向的开发岗位要多一些，所以目前学习前端开发不失为一种明智的选择。其实对于开发岗位来说，更重要的是思考问题的角度和快速解决问题的能力，在精通一门开发语言后，再转另一门就容易多了。

接下来，让我们开始一起了解 Angular 的前世今生吧！

1.1.1 AngularJS 1.x 的诞生

AngularJS 1.x 是一款开源的 JavaScript 库，由 Misko Hevery 和 Adam Abrons 在 2009 年用业余时间创建。起初的项目名字叫 GetAngular，并以 GetAngular.com 注册了网站。由于注册用户过少，两人于是放弃了商业计划并将 Angular 开源了。后来 AngularJS 的发展越来越壮大，最终被 Google 公司所收购，之后就由 Google 负责运营及推广并广泛运用于多款产品当中。

1.1.2 快速发展的 Angular

从 2009 年的 AngularJS 1 到现在的 Angular7，它的发展十分迅速。其中 AngularJS 1.x 与 Angular2 是两个完全不同的框架，需要重新学习，但是从 Angular2 到 Angular7 的框架变化不算太大，还是不难学习的。通过图 1.1 可以看出，Angular 在全球范围内还是拥有着相当大的市场，因为它改过一次名字，所以我们把 AngularJS 的搜索结果一起加进来进行比较。

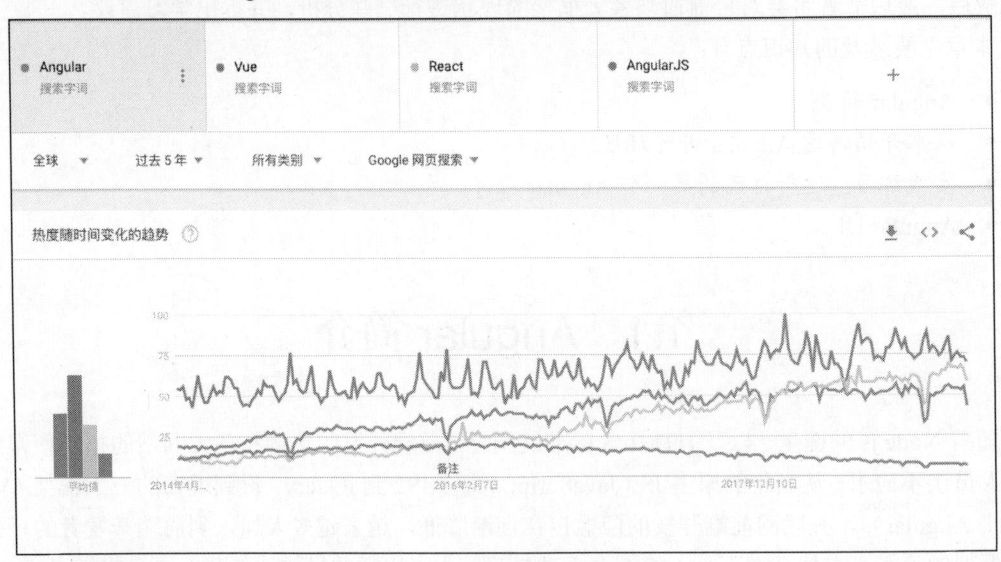

图 1.1 三大框架热度趋势图

下面大概介绍一下从 Angular JS1.x 到 Angular7 的变化。

- **Angular JS1 → Angular2**。Angular JS1 到 Angular2 拆分成了两个完全不同的框架，完全被推翻重写了。Angular JS1 未来将会停止更新。主要区别有：Angular JS1 中的控制器概念已被淘汰，Angular 2 已经改变为基于组件的用户界面。Angular JS1 使用 JavaScript 构建应用程序，Angular 2 使用 TypeScript。
- **Angular2 → Angular4**。Angular 没有发布 3，而是直接从 Angular2 跳到 Angular4。其原

因是@angular/router 已经发布了 v3.3.0，为了在发布新版时让核心库版本号对齐，所以 Angular 框架决定跳过了 3 直接命名为 Angular4，如图 1.2 所示。在 Angular4 中，捆绑文件大小减少了 60%，Angular v4.0 与新版本 TypeScript 2.1 和 TypeScript 2.2 兼容。这有助于更好地进行类型检查。

- **Angular 4 → Angular 5**。这次升级主要是带来了一些新的功能、修正了程序中的错误（bug）、减少打包文件的大小、加快启动的速度等，代码改动比较明显的是 HTTPClient 替代了之前的网络请求库。
- **Angular 5 → Angular 6**。这次的改动不算太大，其中有一些 import 的路径需要更改、用 angular.json 替代 angular-cli.json 和弃用<template>等。
- **Angular 6 → Angular 7**。这次的改动主要完善了 CLI 提示、支持虚拟滚动（基于列表的可见部分从 DOM 中加载和卸载元素，提高应用程序的性能）、支持拖放功能、第三方依赖更新（TypeScript、RxJS、Node10）等。

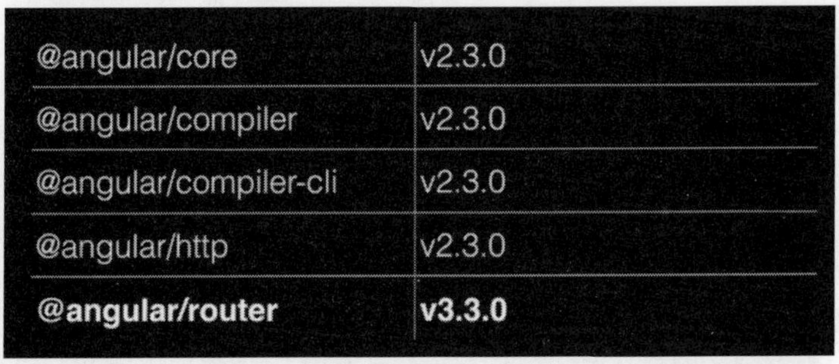

@angular/core	v2.3.0
@angular/compiler	v2.3.0
@angular/compiler-cli	v2.3.0
@angular/http	v2.3.0
@angular/router	**v3.3.0**

图 1.2　@angular/router

1.1.3　三分天下的前端框架

从以前的 jQuery 一统天下，到现在的 Angular、Vue、React 三分天下。既然这三大框架流行了起来，我们也有必要了解它们三者之间的区别，见表 1.1。

表 1.1　三大框架对比

	Angular	Vue	React
组织方式	MVC	模块化	模块化
路由	静态路由	动态路由	动态路由
模板能力	强大	自由	自由
数据绑定	双向绑定	双向绑定	单向绑定
自由度	较小	较大	大

从列表来看，Angular 功能强大，给我们提供了很多东西，但是自由度较小。Vue 早期开发的灵感来源于 AngularJS，一些语法与它很相似，但是更加灵活与简单，适合中小型应用。React 的 JSX 模板相当于 JS，写起来自由度非常大，相当于原生的 JS，对于需要个性化需求的中型应用更加适合。

1.1.4 未来的选择

对于现在前端的情况来说，数量众多的框架让开发者心生畏惧，觉得每天都在出新的东西，永远学不完，心里十分焦虑。程序员应该是框架的主人，我们为了提高效率、解决某些问题才会去选择某框架，但是千万不能成为框架的奴隶。新框架那么多，出一个学一个，什么都会一点，什么都不精。只要拥有扎实的基础知识，对所学框架进行深入的理解，就像学编程语言一样，精通了一门后再学其他语言是很快的。

1.2 搭建开发环境

本节将介绍开发环境的搭建。在环境搭建过程中，我们将会以操作系统用户比例最多的Windows操作系统环境进行演示。由于软件安装到Windows、Mac OS、Linux中的基本过程一致，所以对于在其他系统中的安装流程，本书就不再一一描述了。

1.2.1 安装 Node.js 和 NPM

Node.js是一个基于Chrome V8引擎的JavaScript运行环境，而NPM（Node Package Manager，即“node包管理器”）是Node.js默认的、以JavaScript编写的软件包管理系统。安装方法十分简单，打开Node.js中文网的网址http://nodejs.cn/download/，选择读者电脑中操作系统版本对应的Node.js安装包版本，单击下载，如图1.3所示。

Windows 系统 node-v10.5.0-x64.msi	Mac 系统 node-v10.5.0.pkg	源代码 node-v10.5.0.tar.gz
Windows 系统 (.msi)	32 位	64 位
Windows 系统 (.zip)	32 位	64 位
Mac 系统 (.pkg)	64 位	
Linux 系统 (x86/x64)	32 位	64 位
Docker 镜像	官方镜像	
全部安装包	阿里云镜像	

图 1.3　Node.js 与 NPM 下载

在安装Node.js过程中，NPM会一同被自动安装到系统中。在命令行中输入npm -v、node -v可以检测是否成功安装，如图1.4所示。

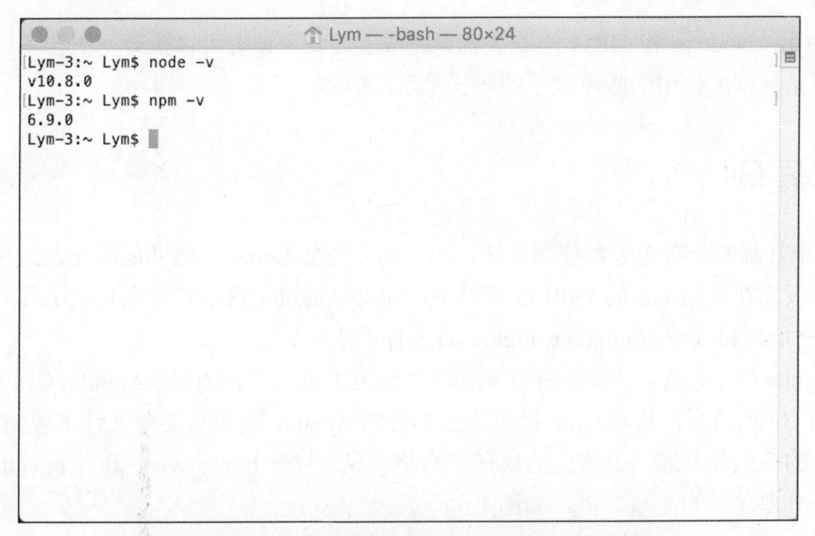

图 1.4　查看当前操作系统安装的 Node.js、NPM 版本

如果读者在使用 NPM 安装第三方库时经常失败，可以选择使用 CNPM。在安装或升级完 Node.js 后运行以下命令可以安装淘宝网提供的 NPM 软件包库的镜像 CNPM，安装命令如下，成功安装后的屏幕输出如图 1.5 所示。

```
npm install -g cnpm --registry=https://registry.npm.taobao.org
```

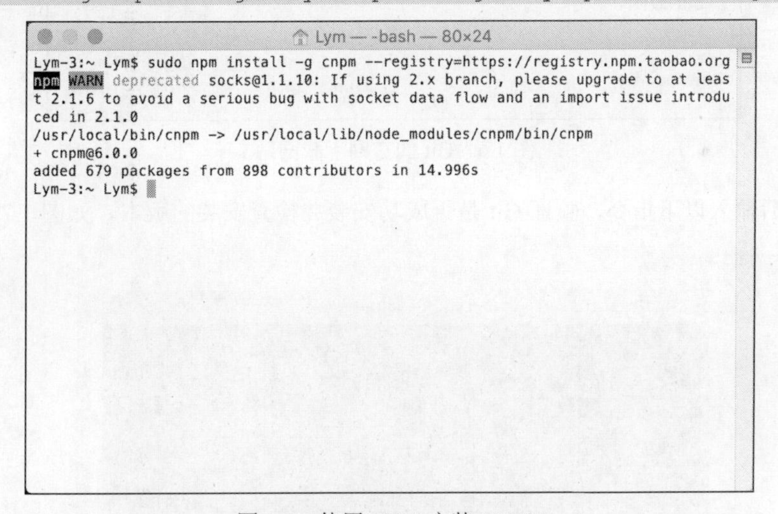

图 1.5　使用 NPM 安装 CNPM

命令成功执行完毕后，将来使用 NPM 命令的地方就都可以用 CNPM 来代替了。本书接下来的内容依然使用 NPM 命令，安装了 CNPM 的读者可以自行替换成 CNPM 命令。CNPM 支持 NPM 中除 publish 之外的所有命令。

提　示

如果使用 npm install 等命令进行安装失败了，Windows 用户可以尝试以管理员身份运行“命令提示符”应用程序，Mac OS、Linux 用户则可以尝试在命令前加上 sudo。

到目前为止，Node.js 和 NPM 的安装就算完成了。最后读者可以检查一下在本地电脑上安装的版本号，尽量保证所使用的版本大于或等于笔者的版本。

1.2.2 安装 Git

Git 是一个开源的分布式版本控制系统，由 Linux 之父 Linus Torvalds 所开发。目前基本所有的开源项目都发布在使用 Git 的 GitHub 网站上，包括 Angular 这个开源项目也上传到了该平台，其 GitHub 的网址为 https://github.com/angular/angular。

使用 Angular CLI 之前，需要在操作系统中安装好 Git。这样当使用 Angular CLI 创建项目时，将会自动调用 Git 的命令，从 GitHub 把对应版本号的 Angular 模板与支持文件下载到本地电脑中。

对于 Git 的安装的首选，当然也是从官方网站获取，打开 https://www.git-scm.com/download/，选择对应的操作系统单击下载即可，如图 1.6 所示。

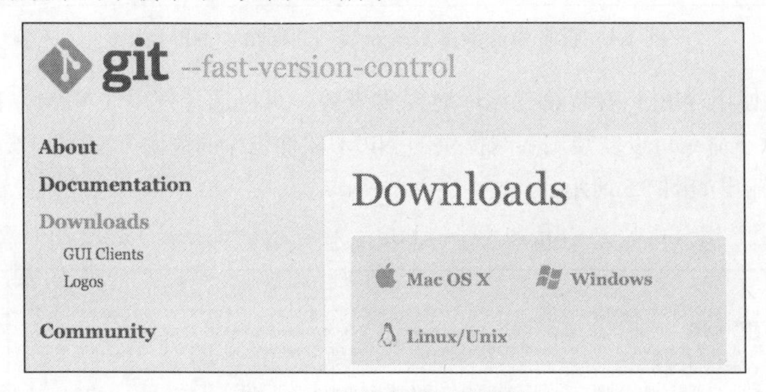

图 1.6 Git 的官网下载网站

安装完毕后输入以下指令，验证 Git 是否成功安装并检查安装的版本，如图 1.7 所示。

```
git --version
```

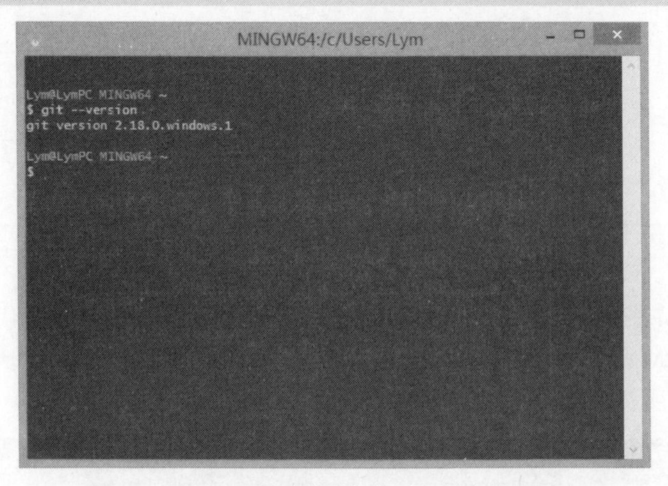

图 1.7 验证 Git 是否成功安装以及被安装的版本

1.2.3　安装 Angular CLI

接下来需要使用刚才成功安装的 NPM 进行 Angular CLI 的安装，Angular CLI 是 Angular 的命令行界面工具，主要用来创建项目、添加文件以及启动服务等功能。使用以下命令进行安装。

```
npm install -g @angular/cli
```

在安装完成后，使用以下命令验证 Angular CLI 是否成功安装以及对应的安装版本，如图 1.8 所示。

```
ng --version
```

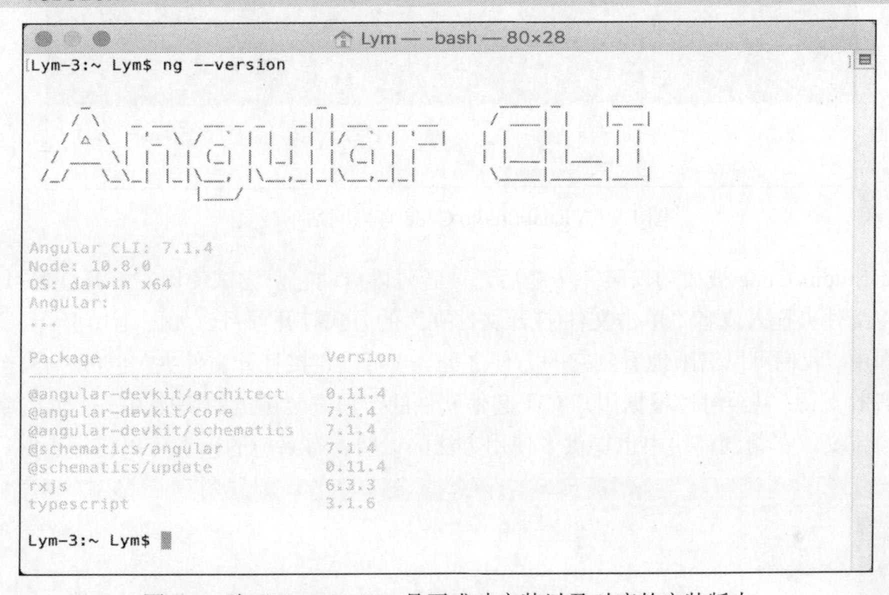

图 1.8　验证 Angular CLI 是否成功安装以及对应的安装版本

1.2.4　开发工具的选择

Visual Studio Code 是由微软公司开发的一个轻量且强大的代码编辑器，不仅免费开源，而且还提供了相当丰富的插件。Angular 所使用的编程语言 TypeScript 也是由微软公司开发的，所以该编辑器对 TypeScript 的支持性也比较强，因此笔者推荐读者使用该编辑器进行开发。如果读者已经习惯了比如 WebStorm、Sublime、Notepad 之类的软件，那可以跳过这一节直接进入下一节。

关于这个编辑器的安装，还是直接到其官方网站 https://code.visualstudio.com/进行下载安装，如图 1.9 所示。

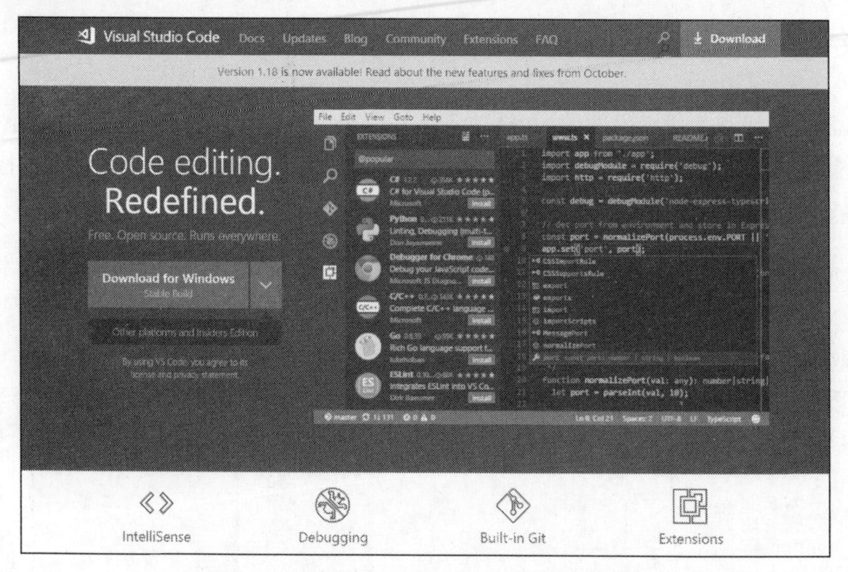

图 1.9　Visual Studio Code 官方网站下载页

Visual Studio Code 被成功安装完毕之后，读者可以启动它，尝试使用它打开任何目录，可以通过直接将文件夹拖入或者“单击文件|打开文件夹”的方式打开项目，如图 1.10 所示。在这个欢迎使用界面中，我们可以清晰地看到它的方便之处。它可以直接打开文件夹；也可以自定义对于各种编程语言的支持；甚至可以根据用户对于其他软件的习惯安装键盘的快捷方式；对于颜色主题也可以进行自定义。在帮助模块中也提供了使用方法的说明，有兴趣的读者可以自行查阅。

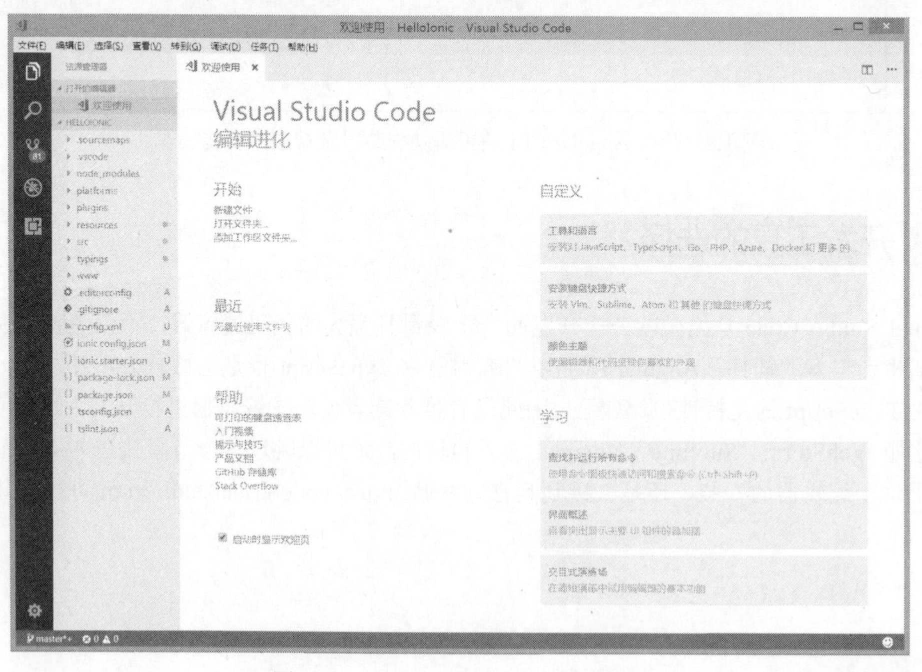

图 1.10　Visual Studio Code 窗口布局

1.2.5　安装 Angular 辅助编码插件

Visual Studio Code 的扩展功能非常强大，可以直接进行搜索和安装，如图 1.11 所示。

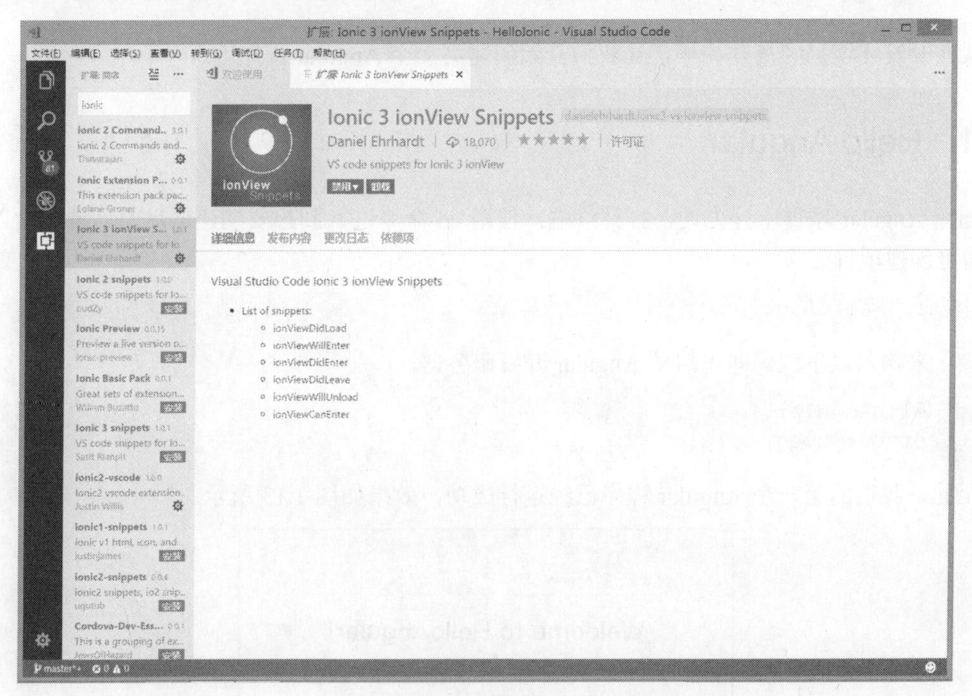

图 1.11　安装 Visual Studio Code 的 Ionic 扩展功能

为了方便 Angular 的开发，读者尽量选择当时最新并且评分比较高的扩展功能进行安装。安装完成后，单击重新加载来刷新页面即可使用。这些扩展功能主要包括了相关的代码提示等功能，方便我们进行开发，但并不是必需的。

在键盘上分别按下 Ctrl+K、Ctrl+T（Mac 上是 Command+K、Command+T）可以进行主题颜色设置，也可以选择安装扩展的主题颜色、扩展的图标等，如图 1.12 所示。笔者为了清晰展示这个界面效果，选择了方便读者观看的浅色主题。

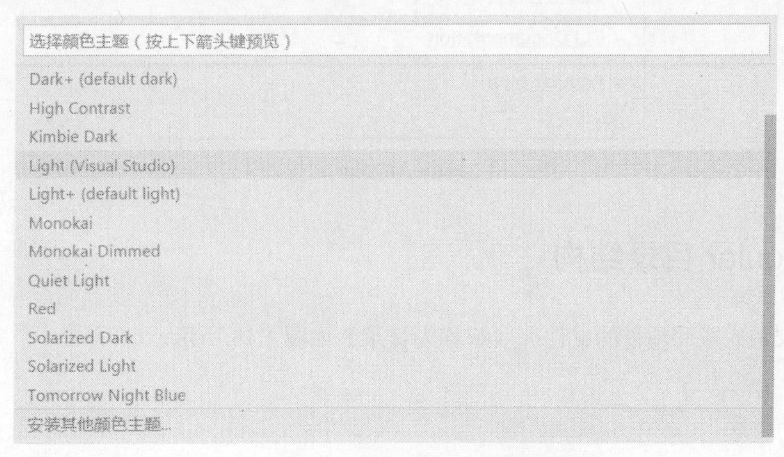

图 1.12　Visual Studio Code 的颜色主题

1.3 实战练习：第一个 Angular 程序

现在开发环境已经搭建完毕，让我们开始创建第一个 Angular 程序吧。

1.3.1 Hello Angular

创建 Angular 项目十分方便，打开终端，使用 cd 命令进入到想要创建项目的目录，输入以下代码即可创建项目。

```
ng new HelloAngular
```

接下来输入以下代码即可启动 Angular 开发服务器。

```
cd HelloAngular
ng serve --open
```

至此，我们的第一个 Angular 程序已经运行成功，效果如图 1.13 所示。

图 1.13　HelloAngular 运行效果

1.3.2 Angular 目录结构

Angular 安装完成后项目的文件夹（或称为目录）如图 1.14 所示。

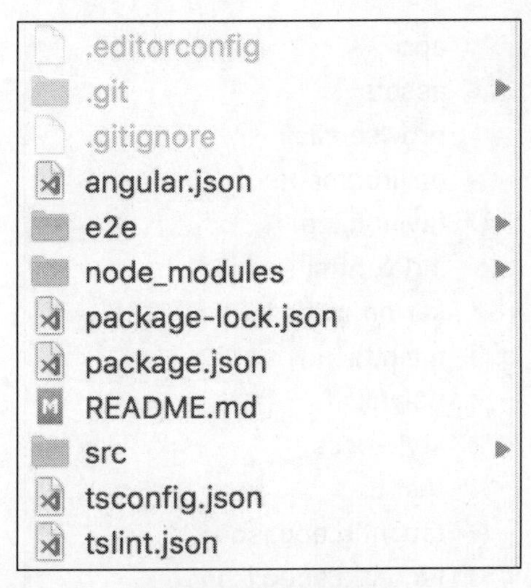

图 1.14 项目的文件夹

从图 1.14 中可以看出，在 HelloAngular 文件夹下生成了很多文件与子文件夹，接下来我们就按照文件的排列顺序进行讲解。

- .editorconfig：编辑器用的配置文件。
- .git/：创建 git 仓库需要的文件夹。
- .gitignore：git 配置文件，会在提交时忽略不需要的文件。
- angular.json：Angular CLI 的配置文件。
- e2e/：端到端（end-to-end）测试所需要的配置文件。
- node_modules/：执行 npm install 后生成的文件夹，会把 package.json 中的第三方模块全部安装在里面。
- package-lock.json：执行 npm install 后生成的文件，锁定了安装时第三方的版本号，上传 git 后可以保证其他人 npm install 时版本号一致。
- package.json：npm 配置文件，列出了项目使用到的第三方模块。
- README.md：项目说明文档，使用 markdown 进行编写。
- src/：主要代码存放的文件夹，之后我们的工作大多都在这个文件夹下完成。
- tsconfig.json：TypeScript 编辑器的配置，用于代码提示。
- tslint.json：使用 TSLint 进行代码检查的配置文件。

Src 文件夹下存放了我们主要的代码，接下来分析它里面的文件。src 文件夹如图 1.15 所示。

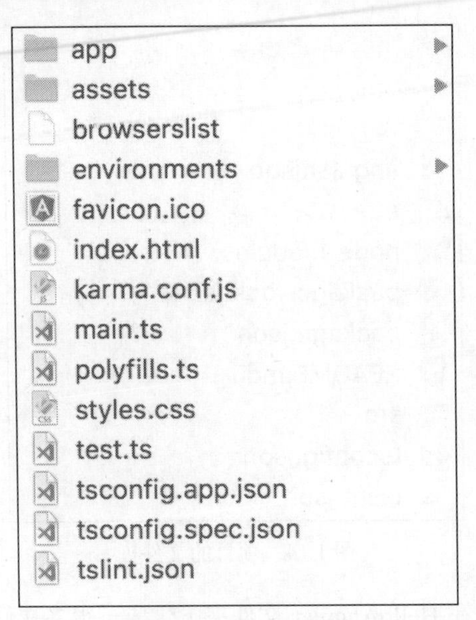

图 1.15 src 文件夹

从图中可以看出，在 HelloAngular 项目文件夹中：

- app/：根组件 APPComponent 所在的位置，也可以对它进行修改。
- assets/：该文件夹下一般存放图片等资源文件。
- browserslist：一个配置文件，用来在不同的前端工具之间共享目标浏览器。
- environments/：环境配置文件，一般用于构建应用时替换文件。比如项目分为开发环境和生产环境等。
- favicon.ico：浏览器上运行时显示的网页图标。
- index.html：主页面的 HTML 文件。大多情况下只需要在根模块处理就行，很少编写这个文件。
- karma.conf.js：Karma 单元测试的配置文件。
- main.ts：应用主入口，根模块就是在这里启动的。
- polyfills.ts：将不同浏览器的支持进行标准化的配置文件。
- styles.css：全局 css 样式。
- test.ts：单元测试入口。
- tsconfig.app.json：应用的 TypeScript 编译器配置文件。
- tsconfig.spec.json：单元测试的 TypeScript 编译器配置文件。
- tslint.json：帮助保持代码风格的一致性的工具。

1.3.3 Angular CLI 详解

使用 ng help 命令可以查看帮助，如图 1.16 所示。

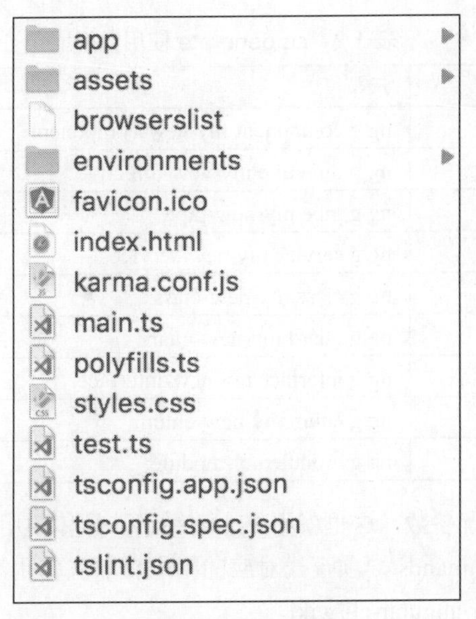

图 1.16　ng help 帮助

- ng add：添加支持库到项目中。
- ng new：新建项目。
- ng generate：新建各种类型的文件。
- ng update：更新程序及依赖的项目。
- ng build：构建程序，默认输出到 dist/。
- ng serve：构建并运行程序，文件更改时可以自动生效。
- ng test：执行单元测试。
- ng e2e：执行 e2e 测试。
- ng lint：执行 TSLint 检查代码。
- ng xi18n：提取国际化信息。
- ng run：运行 Architect 目标。
- ng eject：暂时禁用，退出应用并输入正确的 webpack 配置和脚本。
- ng config：获取/设置配置项。
- ng help：获取 Angular CLI 帮助列表。
- ng version：获取 Angular CLI 版本号。
- ng doc：打开所搜关键字的 Angular API 文档，比如搜索 component，就输入 ng doc component。
- ng new project-name：新建项目，我们一开始创建的 HelloAngular 就是用这个命令创造的。

这里详细讲解 ng generate 的用法，它可以简写为 ng g。后面的参数是类型与文件名，比如 component 是组件、directive 是指令，可以参考表 1.2 的用法快速进行文件的创建。

表 1.2 ng generate 使用详解

基本用法	示范
Component	ng g component my-new-component
Directive	ng g directive my-new-directive
Pipe	ng g pipe my-new-pipe
Service	ng g service my-new-service
Class	ng g class my-new-class
Guard	ng g guard my-new-guard
Interface	ng g interface my-new-interface
Enum	ng g enum my-new-enum
Module	ng g module my-module

这些命令中还有很多配置参数，这里就不一一列举了。具体的用法可以根据上文所列的名称，到以下网址的 Additional Commands 一项查找对应的指令获取具体用法：

https://github.com/angular/angular-cli/wiki。

1.3.4 如何学习 Angular

Angular 的学习曲线比较陡峭，刚入门的时候可能会觉得比较吃力。笔者的建议是跟着本书或者网上的 Demo（示例）多动手练习编写程序，编写得多了自然就熟练了。还有就是要多花费点精力，将 Angular 的架构彻底消化吸收，知道它的各个模块是如何配合工作的。这样在学习组件、指令、模板等知识点时就不会出现理不清思路的情况。

1.4 Angular UI 库

Angular 给开发者提供了方便的框架来编写业务逻辑，本节就来介绍现在 Angular 中较为流行的 UI（用户界面）框架。一个好的 UI 框架，不只是提供好看的页面风格，而是使开发效率事半功倍，让开发者把注意力集中于核心业务逻辑的开发，而不需要在 UI、CSS 上耗费过多精力。下面就开始介绍比较常见的 UI 框架。

1.4.1 NG-ZORRO

Ant Design 作为一门优秀的程序设计语言，经历过多年的迭代和发展，它的 UI 设计已经有了一套别具一格的风格，截至目前在 GitHub 已经有 40000 余颗 Star（星）了，可以说是 React 开发者手中的神兵利器。作为 Ant Design 的 Angular 实现，NG-ZORRO 不仅继承了 Ant Design 的独特思想和极致体验，同时也结合了 Angular 框架的优点和特性。而且该框架由阿里巴巴公司的前端团队开发，中文文档丰富，学习方便。本书在之后的章节中，也将使用该框架进行示例项目的开发。

官方网站：https://ng.ant.design，如图 1.17 所示。

图 1.17　NG-ZORRO 官方网站

1.4.2　Angular Material

Angular Material 可以算是官方的 UI 框架了，该框架的设计思想依托于 Google 公司设计的 Material Design（MD，材料设计语言）。MD 这种设计风格，并不局限于 WEB 开发，它在安卓等设备上也同样被广泛运用。由于是官方开发的，所以它在设计上会与 Angular 契合度更好。缺点则是官方文档为纯英文，而且服务器在国外因而访问特别慢。综合来说，还是一门十分优秀的 UI 框架，感兴趣的读者可以自行了解一下。

官方网站：https://material.angular.io，如图 1.18 所示。

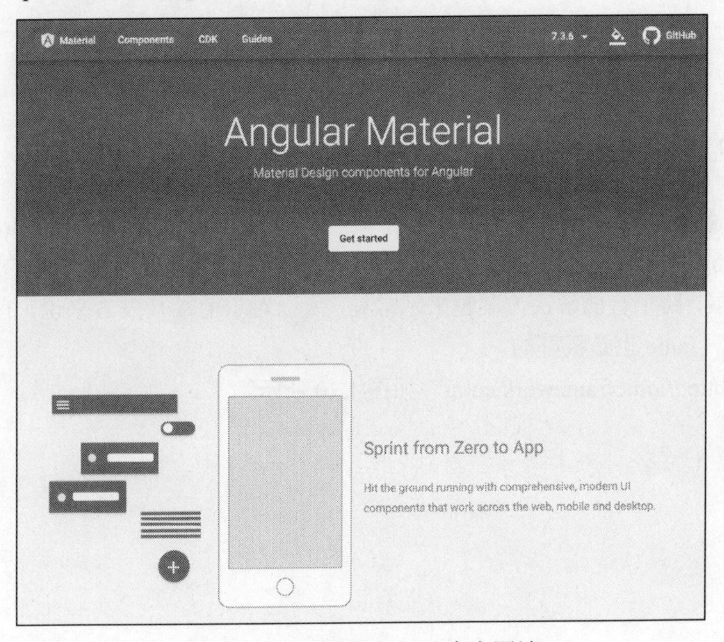

图 1.18　Angular Material 官方网站

1.4.3 ng-bootstrap

Bootstrap 是前端开发界的老牌 UI 框架了，它由 Twitter 的设计师 Mark Otto 和 Jacob Thornton 合作开发。这个框架从初代开始到现在已经出到了 Bootstrap4 了，从这种不间断的维护可以看出还是有自己的固定受众的。为了方便在 Angular 中使用，Bootstrap 推出了专属于 Angular 的 ng-bootstrap，以更好地与之契合。由于 ng-bootstrap 关注度不是很高，因此并没有像 Bootstrap 一样有人将官方文档翻译为中文，对于英文不是很好的开发者，在学习过程中可能会造成一些麻烦。当然，如果读者非常喜欢 Bootstrap 的设计风格，选择它也未尝不可。

官方网站：https://ng-bootstrap.github.io，如图 1.19 所示。

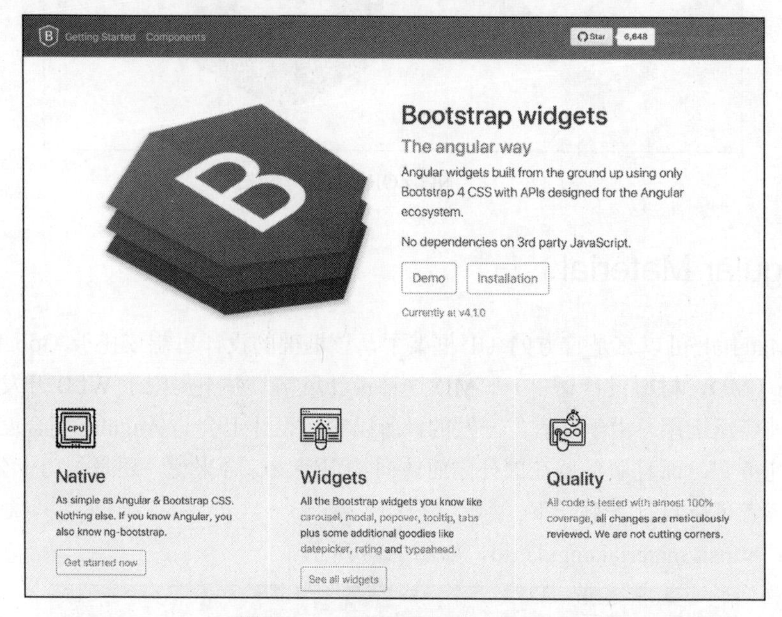

图 1.19　ng-bootstrap 官方网站

1.4.4 Ionic

Ionic 是为了解决移动端应用开发而诞生的一个 UI 框架，实际上 Ionic = Cordova + Angular + Ionic CSS，由 Angular 作为主要框架，Ionic 是 UI 样式，而 Cordova 用来生成 iOS/Android 项目。随着大前端的发展，跨平台的解决方案越来越常见，所以对于前端开发者来说，有机会开发移动端应用时，尝试一下 Ionic 也是极好的。

官方网站：http://ionicframework.com/，如图 1.20 所示。

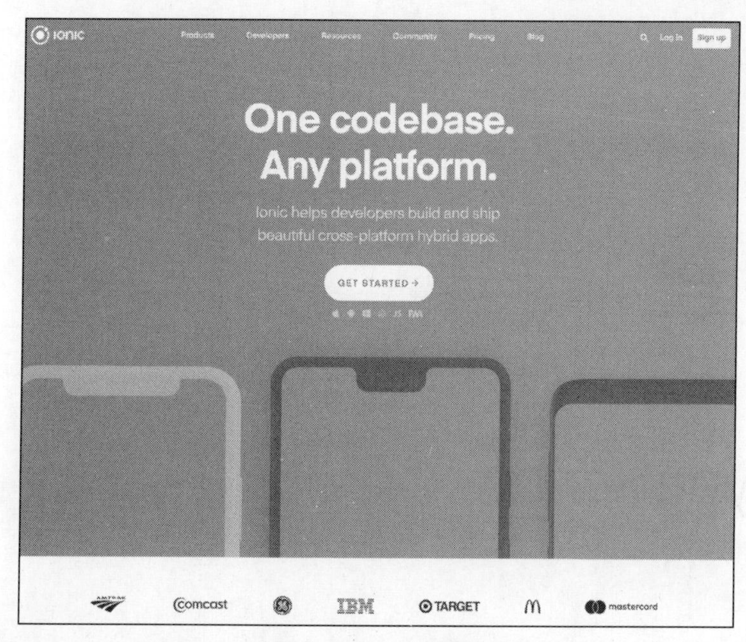

图 1.20　Ionic 官方网站

1.5　小结

本章介绍了 Angular 的诞生与发展，以及如何开始自己第一个程序 HelloAngular。此外，本章将 Angular 目录结构与 Angular CLI 列举出来方便读者查阅，在后续的章节中，读者将会一一学习和接触到这些内容。通过第 1 章的学习，读者应该已经发现本书比较注重实践，而不会在介绍框架上占用两三个章节。Linus Torvalds 说过一句话“**Talk Is Cheap, Show Me The Code**”（**能说算不上什么，有本事就把你的代码给我看看**）。笔者同样认为，比起大篇幅的概念介绍，更应该多实践，毕竟比起纸上谈兵，动手编写代码才是提升程序开发技能最快的方法。

第2章

初识 TypeScript

通过最近的 Stack Overflow 开发者调查以及年度 RedMonk 编程语言排名显示，TypeScript 的人气变得越来越高。为什么它会如此火热呢？首先 TypeScript 依托于微软的官方支持，而且它完全兼容 JavaScript，作为 Angular 官方首推的语言，使用起来肯定比原生 JavaScript 更胜一筹。总的来说，TypeScript 的学习十分有必要，希望读者通过本章的学习可以加深对 TypeScript 的理解。

本章主要涉及的知识点有：

- TypeScript 简介
- 数据类型
- 函数
- 类

2.1 TypeScript 简介

正如这门编程语言的名字，作为 JavaScript 的超集，TypeScript 添加了可选的静态类型和基于类的面向对象编程。这门编程语言是由微软公司的 Anders Hejlsberg 主导开发的，并于 2013 年 6 月 19 日正式发布。TypeScript 是一门开源的编程语言，众多的开发者都在为完善这个项目做出自己的贡献，其项目地址为 https://github.com/Microsoft/TypeScript。

2.1.1 动态类型语言与静态类型语言

我们先分别介绍一下基本概念。

（1）动态类型语言

动态类型是指在运行期才进行数据类型的检查。主要优点在于可以少写很多类型声明代码，编程会更自由并且程序更易于阅读。JavaScript 就是一门动态类型语言。

（2）静态类型语言

静态类型语言在编译期就会进行数据类型的检查。主要优点在于：因为类型声明是强制的，所以使得 IDE（集成开发环境）有很强的代码感知能力，能在早期发现一些问题，方便程序的调试。TypeScript 就是一门静态类型语言。

总的来说，动态类型语言和静态类型语言各有各的优势。其实它们之间主要的区别就是适合的使用场景不同，没有必要在哪种更好上进行争论。

2.1.2　开发环境搭建

在运行本章节的示例代码前，需要先搭建 TypeScript 的运行环境，或者在网上找一个能在线运行 TypeScript 的网站来运行代码。

它的安装方式十分简单，还是使用 NPM 进行安装。

```
npm install -g typescript
npm install -g ts-node
```

其中前者是安装 TypeScript，后者则是安装 TypeScript 的 Node 运行环境。安装完成后可以输入以下指令检查版本号，如图 2.1 所示。

```
tsc --version
ts-node --version
```

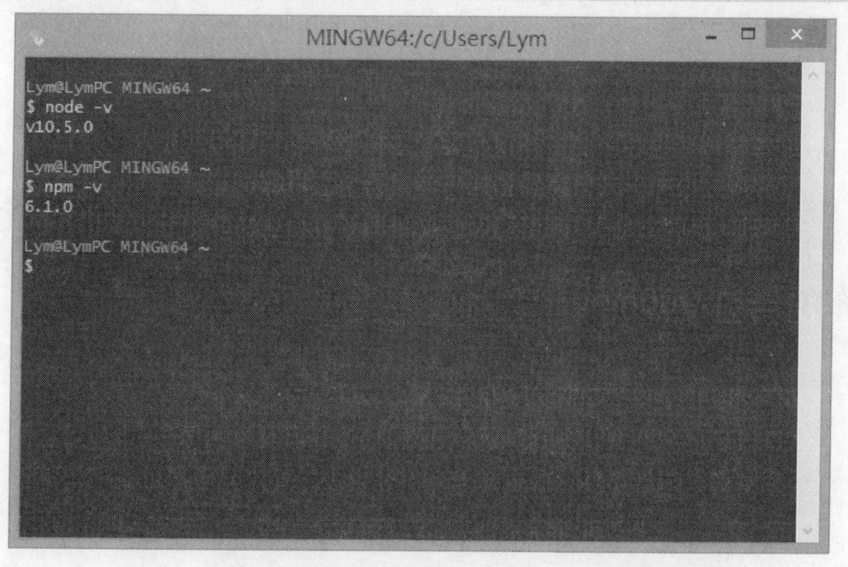

图 2.1　查看当前操作系统安装的 TypeScript、ts-node 版本

运行代码的方式也很简单，新建一个 test.ts 文件，并输入一行测试代码。

```
console.log('Hello TypeScript');
```

之后使用命令行进入到 test.ts 文件目录并使用 ts-node 运行该文件。

```
ts-node test.ts
```

运行成功后结果如图 2.2 所示。

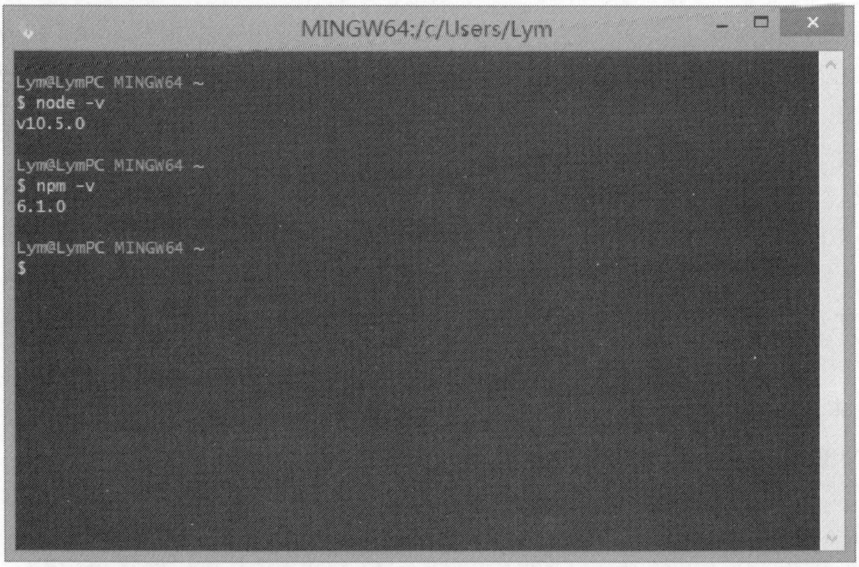

```
MINGW64:/c/Users/Lym
Lym@LymPC MINGW64 ~
$ node -v
v10.5.0

Lym@LymPC MINGW64 ~
$ npm -v
6.1.0

Lym@LymPC MINGW64 ~
$
```

图 2.2　查看当前操作系统安装的 TypeScript、ts-node 版本

> **提　示**
>
> ts-node 工具主要依赖的是 Node.js，让代码可以脱离浏览器运行，它自己的作用则是将 TypeScript 代码编译为 JavaScript 代码。

2.2　数据类型

本节将介绍 TypeScript 中常见的数据类型，并会在每一小节给出相关的示例代码。

2.2.1　布尔类型

与大多数语言一样，TypeScript 中布尔类型的值为 true/false。

【示例 2-1】定义一个布尔类型变量，赋值为 false 并输出。

```
let isDog: boolean = false;
console.log(isDog);

// 输出结果：false
```

2.2.2　数字类型

TypeScript 中数字类型都是浮点数，所以整数可以直接与带小数点的数字进行运算。

【示例 2-2】定义两个数字类型变量，分别赋值整数与浮点数进行相加。

```
let num1: number = 10;
let num2: number = 5.5;
console.log(num1 + num2);

// 输出结果：15.5
```

2.2.3 字符串类型

TypeScript 中字符串类型使用单引号与双引号来表示，并且支持 ES6（即 ECMAScript 6）的反引号来进行操作。

【示例 2-3】分别用单引号和双引号定义两个字符串类型变量，最后使用反引号拼接输出。

```
let str1: string = 'Hello';
let str2: string = "TypeScript";
console.log(`${str1} ${str2}!`);

// 输出结果：Hello TypeScript!
```

2.2.4 数组类型与元组类型

TypeScript 中定义数组的方式有两种，元组类型实质上也是数组类型，只是允许数组添加不同类型的值。

【示例 2-4】分别使用不同的方式初始化一个新的数组，并示范如何创建元组类型数组。

```
// 元素类型后接中括号
let arr1: number[] = [1, 2, 3];

// 使用数组泛型，Array<元素类型>
let arr2: Array<number> = [1, 2, 3];

// 定义元组类型
let arr3: [string, number];

arr3 = ['Hello', 10]; // 正确的赋值方式
arr3 = [10, 'Hello']; // 错误的赋值方式
```

2.2.5 枚举类型

为了防止代码出现过多的魔法数字（在编程领域指的是莫名其妙出现的数字，而数字的意义必须通过详细阅读才能推断出来，一般魔法数字都是需要使用枚举变量来替换的）等情况，所以枚举类型的存在是十分必要的。与其他语言类似，TypeScript 中默认情况下从 0 开始编号，也可以手动指定成员数值。

【示例 2-5】定义两组枚举，并使用不同方式指定数值。

```
// 指定第一个
enum Color {Red = 1, Green, Blue}
console.log('Red=' + Color.Red);
```

```
console.log('Green=' + Color.Green);
console.log('Blue=' + Color.Blue);

// 全部指定
enum Animal {Dog = 2, Cat = 4, Bird = 5}
console.log('Dog=' + Animal.Dog);
console.log('Cat=' + Animal.Cat);
console.log('Bird=' + Animal.Bird);

// 输出结果：Red=1  Green=2  Blue=3    Dog=2 Cat=4 Bird=5
```

2.2.6 any 类型

Any 类型实际上相当于移除了类型检查，它允许我们像 JavaScript 一样给它赋值任意类型。

【示例 2-6】定义一个 any 类型变量，先赋值为数字类型的 18，后赋值为字符串类型的 eighteen，最后进行输出。

```
let age: any = 18;
age = 'eighteen';
console.log(age);

// 输出结果：eighteen
```

2.2.7 void 类型

void 类型的意思就是没有类型。对于变量来说这个类型只能赋值 undefined 和 null，而函数返回值为空通常可以省略，所以用处不是很大。

【示例 2-7】示范 void 类型的使用。

```
// undefined
let unusable: void = undefined;

// 返回值为空
function someFunction(): void {}
// 通常省略为
function someFunction() {}
```

2.2.8 null 与 undefined 类型

null 与 undefined 类型其实都各自有一个特殊的值，分别是 null 和 undefined（注意：类型名和值同名）。null 与 undefined 两种类型和 void 类型一样作用并不大。一般而言，null 和 undefined 是所有类型的子类型，可以作为值赋值给其他类型，除非我们明确指定了--strictNullChecks 标记，才能强制 null 和 undefined 值只能赋值给自己的这种类型，即 null 作为值赋值给 null 类型，undefined 作为值赋值给 undefined 类型。

2.3 函 数

在 TypeScript 中函数这个概念十分重要，它在 JavaScript 的函数基础上增加了更多便捷的功能，对开发效率的提升十分显著。

2.3.1 函数的使用

与 JavaScript 一样，TypeScript 可以创建带名字的函数，也可以创建匿名函数。我们分别对这两种方式进行演示。

【示例 2-8】示范 void 类型的使用。

```
function addNumber(a: number, b: number) {
  return a + b;
}
console.log(addNumber(1, 2));

// 输出结果：3

let addString = function (a: string, b: string) {
  return a + b;
};
console.log(addString('Hello', 'TypeScript'));

// 输出结果：HelloTypeScript
```

2.3.2 构造函数

构造函数是一种特殊的函数。它主要用于创建对象时初始化对象，常与 new 运算符一起使用。TypeScript 的构造函数用关键字 constructor 来实现，可以通过 this 来访问当前类内的属性和方法。

【示例 2-9】创建一个简单的构造函数，并用它初始化一个新对象进行输出。

```
class Student {        // 定义 Student 类
  name: string;  // 定义类的属性 name
  age: number;   // 定义类的属性 age
  constructor(name: string, age: number) { //定义构造函数
      this.name = name;
      this.age = age;
  }
}

let stu = new Student('LiSi', 18);
console.log('name=' + stu.name + ' age=' + stu.age);

// 输出结果：name=Wang Cai age=2
```

2.3.3 可选参数

TypeScript 支持给函数设置可选参数，只要在参数后面增加问号标识即可实现。

【示例 2-10】新建一个带有可选参数的函数，并进行可选传值测试。

```
function add(a: number, b?: number) {
  if (b) {
    return a + b;
  } else {
    return a;
  }
}

console.log(add(3, 2));

// 输出结果：5

console.log(add(3));

// 输出结果：3
```

提 示

使用可选参数时，其中有一个规则读者需要注意一下，就是必选参数不能位于可选参数后面，这种错误的用法将会导致编译报错。

2.3.4 默认参数

TypeScript 同样支持给函数设置默认参数，只要在参数声明后用赋值运算符（等号）进行赋值即可。

【示例 2-11】新建一个带有默认参数的函数，并进行默认传值测试。

```
function add(a: number, b: number = 5) {
  if (b) {
    return a + b;
  } else {
    return a;
  }
}

console.log(add(3, 2));

// 输出结果：5

console.log(add(3));

// 输出结果：8
```

2.3.5　箭头函数

JavaScript 中 this 的作用域是一个常见的问题，看下面这个例子。

```
const Person = {
  'name': 'LiSi',
  'printName': function () {
    let fun = function () {
      return '姓名：' + this.name;
    };
    return fun();
  }
};
console.log(Person.printName());

// 输出结果：姓名：undefined
```

遇到这种问题通常会用以下方法进行解决，声明一个变量 self 在函数外部先绑定上正确的 this，在函数内部再通过 self 调用 name 属性。

```
const Person = {
  'name': 'LiSi',
  'printName': function () {
    let self = this;
    let fun = function () {
      return '姓名：' + self.name;
    };
    return fun();
  }
};
console.log(Person.printName());

// 输出结果：姓名：LiSi
```

箭头函数提供了另一种方便的解决方案，它内部的 this 是词法作用域，所以我们不需要再进行多余的操作，看一下下面这个例子。

【示例 2-12】使用箭头函数解决 this 作用域问题。

```
const Person = {
  'name': 'LiSi',
  'printName': function () {
    let fun = () => {
      return '姓名：' + this.name;
    };
    return fun();
  }
};
console.log(Person.printName());

// 输出结果：姓名：LiSi
```

使用这种方式不仅使代码更加精简清晰、便于阅读，而且也规避了一些可能会出现的错误。

2.4 类

不同于 JavaScript 使用函数和基于原型的继承，TypeScript 中是基于类的继承并且对象是由类构建出来的。

2.4.1 属性与方法

TypeScript 中创建属性的方法与 JavaScript 类似。一般变量使用 let、常量使用 const 进行创建。变量类型在变量名后面加冒号来声明，而类型声明则可以省略。TypeScript 中方法则需要创建在类中，默认为 public，返回值是在方法名后加冒号来进行声明，如果无返回值可以省略或使用 void。

【示例 2-13】属性与方法的使用示范。

```
// 属性示例
let name1: string = 'LiSi';
let name2 = 'LiSi';
const name3 = 'LiSi';

// 方法示例
class User {
  getUserName(): string {
    return 'LiSi';
  }
}
let user = new User();
console.log(user.getUserName());
// 输出结果：LiSi
```

2.4.2 类的继承

TypeScript 中类的继承是通过 extends 关键字来实现的，派生类通常被称为子类，基类通常被称为父类。

【示例 2-14】简单示范 TypeScript 中类的继承的使用方法。

```
class Person {
  name: string;
  constructor(name: string) {
      this.name = name;
  }
}

class Student extends Person {
  constructor(name: string) {
      super(name);
  }
}

let stu = new Student('LiSi');
```

```
console.log(stu.name);

// 输出结果：LiSi
```

【代码分析】在这个例子中 Student 类继承了 Person 类，并在构造函数中调用了父类的构造函数，最终输出了父类的属性 name。

2.4.3 访问权限修饰符

与众多的编程语言一样，TypeScript 有自己的访问权限修饰符 public、private、protected，接下来分别举例讲解一下不同访问权限修饰符的用法。

1. public

使用 public 修饰（或定义）的成员是公共的（即公有的，公共的），可以在外部访问到。在 TypeScript 中，成员都默认为 public。

【示例 2-15】

```
class Student {
  public name: string;
  constructor(name: string) {
      this.name = name;
  }
}
let stu = new Student("LiSi");
console.log(stu.name);

// 输出结果：LiSi
```

2. private

使用 private 修饰（或定义）的成员是私有的，不可以在类的外部访问到。

【示例 2-16】

```
class Student {
  private name: string;
  constructor(name: string) {
      this.name = name;
  }
}
let stu = new Student("LiSi");
console.log(stu.name);

// 输出结果：错误：'name'是私有属性
```

3. protected

使用 protected 修饰（或定义）的成员是受保护的，只可以在派生类访问到。

【示例 2-17】把之前类所继承的例子做了修改，增加了一个 protected 修饰符，根据输出结果可以看到仍然可以访问。

```
class Person {
  protected name: string;
```

```
  constructor(name: string) {
      this.name = name;
  }
}

class Student extends Person {
  constructor(name: string) {
      super(name);
  }
}

let stu = new Student('LiSi');
console.log(stu.name);

// 输出结果：LiSi
```

2.5 小 结

本章首先对 TypeScript 做了简单的介绍，并对现阶段动态类型语言与静态类型语言进行了对比，之后搭建了开发环境并对其常见的知识点作了讲解。最后需要明确一点，TypeScript 的出现并不是要完全取代 JavaScript，它们两个虽然同源但各有各的特色，学了 TypeScript 之后并不需要完全放弃 JavaScript。最后，在掌握了 TypeScript 的相关基础知识后，基本上使用这门语言进行 Angular 应用开发已经不成问题了，相信熟悉 JavaScript 的读者一定能很快地掌握本章节的内容。

第3章 指令

本章将介绍使用 Angular 指令的基础知识。Angular 拥有一套完整、可扩展、用来帮助 Web 应用开发的指令集机制。

指令的本质是“当关联的 HTML 结构进入编译阶段时应该执行的操作”，它只是一个当编译器编译到相关 DOM 时需要执行的函数，可以被写在 HTML 元素的名称、属性、CSS 类名和注释里，比如组件实际上就是指令的一种。

本章主要涉及的知识点有：

- 指令的分类
- 内置指令
- 实战练习：自定义结构型指令

3.1 指令的分类

Angular 将指令分为三种类型。

- 组件
- 结构型指令
- 属性型指令

3.1.1 组件

Angular 中的组件以及组件化是重要的知识点，相关内容会在第 4 章进行详细讲解。

3.1.2 结构型指令

结构型指令可以改变 DOM 的布局。当浏览器启动、开始解析 HTML 时，DOM 元素上的指令属性就会跟其他属性一样被解析，也就是说当一个 Angular 应用启动时，Angular 编译器就会遍历 DOM 树来解析 HTML，寻找这些指令属性函数，在一个 DOM 元素上找到一个或多个这样的指令属性函数，它们就会被收集起来、排序，然后按照优先级顺序被执行。

3.1.3 属性型指令

属性型指令是改变元素、组件或其他指令的外观和行为的指令。属性型指令至少需要一个带有 @Directive 装饰器的控制器类。该装饰器指定了一个用于标识属性的选择器。

3.2 内置指令

Angular 通过内置属性型指令来扩展 HTML 控件属性，这种内置属性型指令的名称带有前缀 ng，它的实质是绑定在 DOM 元素上的函数。在该函数内部可以操作 DOM、调用方法、定义行为、绑定控制器等。Angular 应用的动态性和响应能力，都要归功于内置属性型指令。比较常见的内置属性型指令有：ngIf、ngFor，此外还有 ngClass、ngStyle、ngSwitch 等多个指令。本书将在它们第一次出现的地方结合示例代码解释其作用。

3.2.1 ngFor

【示例 3-1】像 for 循环一样，可以重复从数组中取值并显示出来。

```
// .ts

this.userInfo = ['张三', '李四', '王五'];

// .html

<div class="ui list" *ngFor="let username of userInfo">
   <div class="item">{{username}}</div>
</div>
```

它的语法是*ngFor="let username of userInfo"，其中 userInfo 是从中取值的数组，username 是每次从中取出来的值。然后在这个标签里面的内容就会重复执行，并通过双向绑定，将 username 显示出来。

3.2.2 ngIf

【示例 3-2】根据条件决定是否显示或隐藏这个元素。

```
// .html
<div *ngIf="false"></div>
<div *ngIf="a > b"></div>
<div *ngIf="username == '张三'"></div>
<div *ngIf="myFunction()"></div>
```

代码说明：

永远不会显示。

当 a 大于 b 的时候显示。

当 username 等于 '张三' 的时候显示。

根据 myFunction()这个函数的返回值决定是否显示。

3.2.3 ngSwitch

【示例 3-3】用于避免在条件复杂的情况下过多地使用 ngIf。

```
// .html

<div class="container" [ngSwitch]="myAge">
    <div *ngSwitchCase="'10'">age = 10</div>
    <div *ngSwitchCase="'20'">age = 20</div>
    <div *ngSwitchDefault="'18'">age = 18</div>
</div>
```

[ngSwitch]先与目标进行绑定，ngSwitchCase 列出每个可能性，ngSwitchDefault 列出默认值。

3.2.4 ngStyle

【示例 3-4】可以使用动态值给指定的 DOM 元素设置 CSS 属性。

```
// .ts
backColor: string = 'red';

// .html
<div [style.color]="yellow">
    你好，世界
</div>
<div [style.background-color]="backColor">
    你好，世界
</div>
<div [style.font-size.px]="20">
    你好，世界
</div>
<div [ngStyle]="{color: 'white', 'background-color': 'blue', 'font-size.px':
'20'}">
    你好，世界
</div>
```

代码说明：

直接设置颜色为 yellow。

设置背景颜色为 backColor，并可以在.ts 文件中对 backColor 的值进行修改。

设置字体大小，需要注意的是只写 font-size 会报错，必须在后面加上.px。当然.em、.%都是可以的。

前面的这三种指令各自都是只设置一个值，而 [ngStyle] 指令可以同时设置多个值，只要使用花括号包住其中的多个设置值的内容即可。需要注意的是连字符（“-”）是不允许出现在对象的键名当中的，如果要使用像 background-color 这类键名时，则需要给它加上单引号。

3.2.5 ngClass

【示例 3-5】动态地设置和改变一个指定 DOM 元素的 CSS 类。

```
// .scss
.bordered {
    border: 1px dashed black;
    background-color: #eee;
}

// .ts
isBordered: boolean = true;

// .html
<div [ngClass]="{bordered: isBordered}">
    是否显示边框
</div>
```

scss 中设置了样式，相当于创建了一个 class="bordered"，在 ts 中新建了一个 isBordered，用于判断是否显示.scss 中的样式。html 中使用 isBordered 作为 bordered 是否要显示出来的判断依据。

3.2.6 ngNonBindable

【示例 3-6】告诉 Angular 不要绑定页面的某个部分。

```
// .html

<div ngNonBindable>
    {{我不会被绑定}}
</div>
```

使用了 ngNonBindable，花括号就会被当成字符串一起显示出来。

3.3 实战练习：自定义结构型指令

结构型指令像其他指令一样，可以把它应用到一个宿主元素上。然后让它对宿主元素及其子元素做点什么，比如 ngIf、ngSwitch 就是这样的结构型指令。

提 示

指令的类名一般拼写成大驼峰形式，如 NgIf，而它的属性名（即指令名）则拼写成小驼峰形式，如 ngIf。指令名前需要带有星号（*）前缀。

3.3.1 星号前缀

在前面的介绍中我们知道，指令前面通常会带有一个星号前缀，但是这个前缀的作用到底是什么呢？

```
<div *ngIf="hero" class="name">{{hero.name}}</div>
```

星号是一个用来简化更复杂语法的“语法糖”（Syntactic Sugar）。从内部实现原理分析举例，*ngIf 指令（或属性）实际上被翻译成一个<ng-template>元素，并用它来包裹宿主元素。

```
<ng-template [ngIf]="user">
  <div>{{user.name}}</div>
</ng-template>
```

所以说*ngIf 指令实际上转移到了<ng-template>标签上，在那里绑定成为一个不带星号的属性 ngIf。<div>中的内容，实际上转移到了内部的<ng-template>标签上。

3.3.2 创建一个结构型指令

接下来我们模仿 ngIf 来实现一个结构型指令，输入以下代码创建一个新项目。

```
ng new directive
cd directive
ng add ng-zorro-antd
```

在新建 Angular 和添加 NG-ZORRO 的时候，命令行可能会提示让我们进行一些选择，如果书中没有特别说明的话，可以按回车键直接选择默认项。接下来新建文件 appIf.directive.ts，并输入以下代码：

```
import { Directive, Input, TemplateRef, ViewContainerRef } from
'@angular/core';

@Directive({ selector: '[appIf]' })
export class AppIfDirective {

  constructor(
    private templateRef: TemplateRef<any>,
    private viewContainer: ViewContainerRef) { }

  @Input() set appIf(condition: boolean) {
    if (condition) {
      this.viewContainer.createEmbeddedView(this.templateRef);
    } else if (!condition) {
      this.viewContainer.clear();
    }
```

```
  }
}
```

【代码解析】通过@Input 以传值方式将 appIf 传进来，并通过判断它的布尔值对组件进行显示或隐藏。至此一个基本的指令就完成了。

3.3.3 响应用户操作

指令完成后，需要进行一次实际操作验证它的可用性。

（1）首先需要在 app.module.ts 中注入该指令。

```
import { BrowserModule } from '@angular/platform-browser';
import { NgModule } from '@angular/core';

import { AppComponent } from './app.component';
import { MatButtonModule } from '@angular/material';
import { AppIfDirective } from './appIf.directive';

@NgModule({
  declarations: [
    AppComponent,
    AppIfDirective
  ],
  imports: [
    BrowserModule,
    MatButtonModule
  ],
  providers: [],
  bootstrap: [AppComponent]
})
export class AppModule { }
```

（2）接下来在 app.component.html 构建用于展示内容的页面，输入以下代码：

```
<!-- 自定义结构型指令 appIf -->

<div style="margin: 15px">

  <h2>自定义结构型指令 appIf</h2>

  <nz-dropdown [nzTrigger]="'click'">
    <button nz-button nz-dropdown nzType="primary">选择性别</button>
    <ul nz-menu>
      <li (click)="changeShow('man')" nz-menu-item>男</li>
      <li (click)="changeShow('woman')" nz-menu-item>女</li>
    </ul>
  </nz-dropdown>

  <p *appIf="isMan">性别：男</p>
  <p *appIf="!isMan">性别：女</p>

</div>
```

【代码解析】这个页面通过一个下拉菜单的单击事件来改变值，最后使用我们自定义的 appIf 指令来控制 p 标签显示的是男还是女。

（3）最后在 app.component.ts 输入以下代码：

```
import { Component } from '@angular/core';

@Component({
  selector: 'app-root',
  templateUrl: './app.component.html',
  styleUrls: ['./app.component.css']
})
export class AppComponent {
  title = 'app';
  isMan = true;

  changeShow(sex: string) {
    if (sex === 'man') {
      this.isMan = true;
    } else {
      this.isMan = false;
    }
  }
}
```

【代码解析】通过 isMan 来控制对 appIf 的传入值，changeShow 方法通过传入的值来修改 isMan 的值。效果如图 3.1 和图 3.2 所示。

图 3.1　isMan 为 true 时显示的文本

图 3.2　isMan 为 false 时显示的文本

到这里这个项目示例就结束了，虽然代码内容不是很多，但是已经完整地展示了如何使用指令来还原 ngIf 所产生的效果。

3.4 小 结

本章的内容是后面学习 Angular 开发的重要知识的基础，理清了指令的基本概念，可以更充分地学习 Angular。为了便于读者理解所学的内置指令，本章最后给出了实战练习，带领读者自己编写一个自定义结构型指令。对于指令的学习，笔者建议一定要多动手进行尝试，以便巩固在本章学到的知识。

第 4 章

使用组件打造你的项目

首先要了解到底什么是组件。组件实质上就是一个特定的功能模块，Angular 这个框架，就是一个由众多组件构成的组件树，每个组件各司其职，共同构造出一个完整的项目。W3C（World Wide Web Consortium，万维网联盟）曾经提出了 Web Component 的标准，要求每个组件包含自己的 HTML、CSS、JavaScript 文件，并且不能对其他组件页面产生影响，这种非侵入的封装方式也被 Angular 所采用。

本章主要涉及的知识点有：

- 组件
- 注解
- 生命周期
- 数据传递
- 实战练习：城市组件

4.1 组 件

如本章开头所说，Angular 是以 Web Component 作为标准进行设计的，并且添加了一些自己特有的功能，我们会在后面详细讲解 Angular 的组件。

4.1.1 组件的组成

组件有一个装饰器，它能接收一个配置对象，Angular 会基于这些信息创建和展示组件及其视图。看一下这段代码：

```
import { Component } from '@angular/core'
```

```
@Component({
selector:   'app',
templateUrl: './app.component.html',
providers:  [ HttpService ]
})
export class AppComponent implements OnInit {
/* . . . */
}
```

首先想创建一个组件，必须要从@angular/core 中引入 Component 装饰器。之后的@Component 这一段叫做注解，这个注解告诉我们：这个类以什么方式与该组件连接。最后的 class 就是类的具体实现，我们的变量、方法都需要在里面进行编写。

4.1.2 组件化思想

组件化这个概念很早之前就已经有了，它不只是存在于前端。在其他的编程语言，甚至在现实生活中往往也能看到组件化的例子。组件化的作用就是把一部分作为一个整体，然后我们可以将它这个整体放到别的地方去使用。组件化需要做到“高内聚低耦合”。

高内聚，内聚标志着组件内部结合的紧密程度，所以组件内的功能尽可能地集中。

低耦合，组件与组件不互相依赖，尽可能独立存在。通常来说耦合度越低，独立性就会越强。

4.2 注 解

在 Angular 中定义一个组件时，一般使用@component 进行注解声明。继续分析前文那段代码：

```
import { Component } from '@angular/core'

@Component({
selector:   'app',
templateUrl: './app.component.html',
providers:  [ HttpService ]
})
export class AppComponent implements OnInit {
/* . . . */
}
```

可以看到 Angular 的注解和 Java 的注解非常相像，接下来对它里面的内容进行分析。

（1）selector

CSS 选择器，它告诉 Angular 在父级 HTML 中查找<app>标签，创建并插入该组件。其中最上级的标签就写在 index.html 当中。

（2）template

组件 HTML 模板的模块相对地址，如果使用 template 来写的话是用“`”这个符号来直接编写 HTML 代码，templateUrl 则是设置 html 代码路径。由于使用 template 来编写代码的话，示例代码行数会特别多，所以本书的 HTML 代码和 TS 代码将会分文件编写。

（3）providers

组件所需服务的依赖注入。

4.3　生命周期

指令和组件的实例有一个生命周期：新建、更新和销毁。

当 Angular 使用构造函数新建一个组件或指令后，就会按下面的顺序在特定时刻调用这些生命周期方法，参见表 4.1。

表 4.1　Angular 生命周期

属性/方法/事件	说明
ngOnChanges()	设置数据绑定输入属性时响应。该方法接受当前和上一属性值的 SimpleChanges 对象。当被绑定的输入属性的值发生变化时调用，首次调用一定会发生在 ngOnInit()之前
ngOnInit()	在 Angular 第一次显示数据绑定和设置指令/组件的输入属性之后，初始化指令/组件。在第一轮 ngOnChanges()完成之后调用，只调用一次
ngDoCheck()	在 Angular 无法或不愿意自己检测时作出反应，进行检测。在每个 Angular 变更检测周期中都会被调用，发生在 ngOnChanges()和 ngOnInit()之后
ngAfterContentInit()	只适用于组件。当把内容投影进组件之后调用。在第一次 ngDoCheck()之后调用，只调用一次
ngAfterContentChecked()	只适用于组件。在每次投影组件内容的变更检测完成之后调用。在 ngAfterContentInit()和每次 ngDoCheck()之后调用
ngAfterViewInit()	只适用于组件。初始化完组件视图及其子视图之后调用。第一次 ngAfterContentChecked()之后调用，只调用一次
ngAfterViewChecked()	只适用于组件。在每次完成组件视图和子视图的变更检测之后调用。在 ngAfterViewInit()和每次 ngAfterContentChecked()之后调用
ngOnDestroy()	当 Angular 每次销毁指令/组件之前调用并进行清理。在这里禁止订阅可观察对象和分离事件处理器，以防内存泄漏。在 Angular 销毁指令/组件之前调用

4.4　数据传递

组件的使用少不了数据传递，我们直接用一个简单的例子讲解一下数据传递。输入以下指令新建一个项目，并加入 ng-zorro 作为我们要使用的 UI 框架。

```
ng new component
cd component
ng add ng-zorro-antd
```

然后用 angular cli 新建一个模态窗（或称为模态弹出式窗口）用的组件。

```
ng g c modal
```

细心的读者可能会发现，除了 html、css、ts 文件以外，还多了一个 spec.ts 文件，这个文件是用来做单元测试的，如果不需要的话，可以在命令后加入--spec=false 来禁止生成这个文件。

4.4.1 数据的输入

【示例 4-1】掌握组件中数据的输入。首先打开 modal.component.ts，输入以下代码：

```
import { Component, OnInit, Input, Output, EventEmitter } from '@angular/core';

@Component({
  selector: 'app-modal',
  templateUrl: './modal.component.html',
  styleUrls: ['./modal.component.css']
})
export class ModalComponent implements OnInit {

  @Input()
  title: string;        // 模态窗的标题
  @Input()
  content: string;      // 模态窗的内容
  @Input()
  okText: string;       // 确定按钮文本
  @Input()
  cancelText: string;   // 取消按钮文本
  @Input()
  isVisible = false;
  @Output()
  isVisibleChange = new EventEmitter(); // dialog 显示状态改变事件
  constructor() { }
  ngOnInit() {
  }

  handleOk(): void {
    this.isVisibleChange.emit(false);
  }

  handleCancel(): void {
    this.isVisibleChange.emit(false);
  }
}
```

在 modal.component.html 输入以下代码：

```
  <nz-modal [(nzVisible)]="isVisible" [nzTitle]="title" [nzOkText]="okText"
[nzCancelText]="cancelText" (nzOnCancel)="handleCancel()"
(nzOnOk)="handleOk()">
    {{content}}
  </nz-modal>
```

【代码解析】nz-modal 是 ng-zorro 提供的一个模态窗，通过把它封装成一个我们的组件，并通过@Input 来接收外部传来的数据，并展示在组件中。其中 isVisible 用来控制是展示还是隐藏，title 控制模态窗的标题，okText 控制确定按钮文本，cancelText 控制取消按钮文本，最后它们会将自己的值传入给 ng-zorro 提供的 API 中。

在 app.component.html、app.component.ts 输入以下代码：

```
// html
<button
  nz-button
  [nzType]="'primary'"
  (click)="showModal()">
    <span>模态窗组件测试</span>
</button>

<app-modal
  [isVisible]="modalIsVisible"
  title="测试标题"
  content="测试内容"
  okText="确定"
  cancelText="取消">
</app-modal>

// ts
import { Component } from '@angular/core';

@Component({
  selector: 'app-root',
  templateUrl: './app.component.html',
  styleUrls: ['./app.component.css']
})
export class AppComponent {
  title = 'component';

  modalIsVisible = false;

  showModal(): void {
    this.modalIsVisible = true;
  }
}
```

【代码解析】app.component 所做的工作主要是创建了一个展示模态窗的按钮，并把自己的值传递给了 modal 组件。

接下来等待自动构建完成，之后单击模态窗组件测试按钮即可看到效果，如图 4.1 所示。

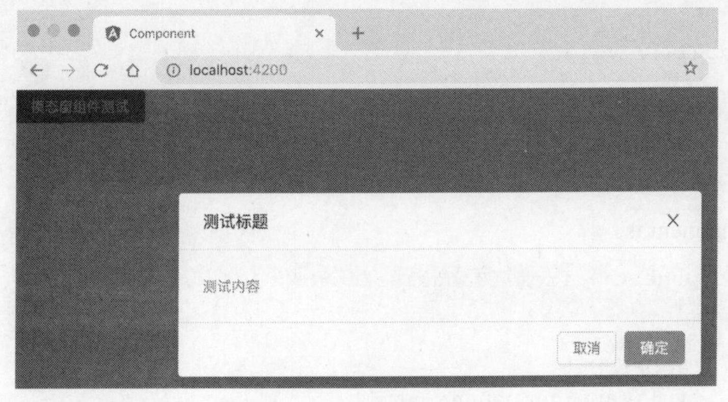

图 4.1　组件的数据输入

从图 4.1 中可以看到,在 app.component.html 中传入的数据已经成功地构建在了 modal 组件中。

4.4.2 数据的输出

【示例 4-2】掌握组件中数据的输出。我们再添加一个功能，根据由组件返回的内容在上级窗口显示组件名。修改 modal.component.ts 代码。

```
import { Component, OnInit, Input, Output, EventEmitter } from '@angular/core';

@Component({
  selector: 'app-modal',
  templateUrl: './modal.component.html',
  styleUrls: ['./modal.component.css']
})
export class ModalComponent implements OnInit {

  @Input()
  title: string; // 模态窗的标题
  @Input()
  content: string;       // 模态窗的内容
  @Input()
  okText: string;        // 确定按钮文本
  @Input()
  cancelText: string;    // 取消按钮文本
  @Input()
  isVisible = false;
  @Output()
  isVisibleChange = new EventEmitter(); // dialog 显示状态改变事件
  @Output()
  clickEvent = new EventEmitter<string>();
  constructor() { }
  ngOnInit() {
  }

  handleOk(): void {
    this.clickEvent.emit('单击确定');
    this.isVisibleChange.emit(false);
  }

  handleCancel(): void {
    this.clickEvent.emit('单击取消');
    this.isVisibleChange.emit(false);
  }
}
```

修改 app.component.ts。

```
import { Component } from '@angular/core';

@Component({
  selector: 'app-root',
  templateUrl: './app.component.html',
  styleUrls: ['./app.component.css']
```

```
})
export class AppComponent {
  title = 'component';
  clickEvent = '';

  modalIsVisible = false;

  showModal(): void {
    this.modalIsVisible = true;
  }

  getClickEvent(eventName: string) {
    this.clickEvent = eventName;
  }

}
```

最后修改 app.component.html。

```
<h2>单击事件：{{clickEvent}}</h2>
<button
  nz-button
  [nzType]="'primary'"
  (click)="showModal()">
    <span>模态窗组件测试</span>
</button>

<app-modal
  [(isVisible)]="modalIsVisible"
  (clickEvent)= "getClickEvent($event)"
  title="测试标题"
  content="测试内容"
  okText="确定"
  cancelText="取消">
</app-modal>
```

分别尝试单击确定和取消，效果如图 4.2 所示。至此，相信读者已经掌握了组件数据的输入和输出了。

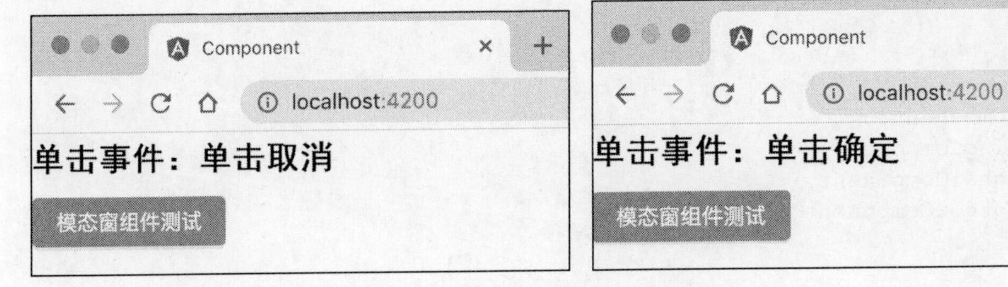

图 4.2　组件的数据输出

4.5 实战练习：城市组件

本节将使用学到的与组件相关的知识，来开发一个小的组件作为练习。读者可能都对一线城市北、上、广非常熟悉，这次的练习就拿它们来做演示数据。我们将使用一个选项卡，来切换选择这三个不同的城市，为这三个城市都设置了对应的组件并展示相对应的数据。

（1）首先新建项目，并导入 NG-ZORRO。

```
ng new city
cd city
ng add ng-zorro-antd
```

（2）新建完成后，通过 Angular-CLI 来给这三个城市创建自己的组件，新建完毕后确认在 app.module.ts 中已经注册好了。

```
ng g component beijing
ng g component shanghai
ng g component guangzhou

// app.module.ts
import { BrowserModule } from '@angular/platform-browser';
import { NgModule } from '@angular/core';

import { AppComponent } from './app.component';
import { NgZorroAntdModule, NZ_I18N, zh_CN } from 'ng-zorro-antd';
import { FormsModule } from '@angular/forms';
import { HttpClientModule } from '@angular/common/http';
import { BrowserAnimationsModule } from
'@angular/platform-browser/animations';
import { registerLocaleData } from '@angular/common';
import zh from '@angular/common/locales/zh';
import { BeijingComponent } from './beijing/beijing.component';
import { ShanghaiComponent } from './shanghai/shanghai.component';
import { GuangzhouComponent } from './guangzhou/guangzhou.component';

registerLocaleData(zh);

@NgModule({
  declarations: [
    AppComponent,
    BeijingComponent,
    ShanghaiComponent,
    GuangzhouComponent
  ],
  imports: [
    BrowserModule,
    NgZorroAntdModule,
    FormsModule,
    HttpClientModule,
    BrowserAnimationsModule
  ],
  providers: [{ provide: NZ_I18N, useValue: zh_CN }],
```

```
  bootstrap: [AppComponent]
})
export class AppModule { }
```

（3）编辑 AppComponent 的代码，创建选项卡，并将选项与所属的组件相对应。代码如下：

```
// app.component.html
<nz-card style="margin: 30px; background-color: #f2f4f5">
  <nz-radio-group [(ngModel)]="radioValue" [nzButtonStyle]="'solid'">
    <label nz-radio-button nzValue="beijing">北京</label>
    <label nz-radio-button nzValue="shanghai">上海</label>
    <label nz-radio-button nzValue="guangzhou">广州</label>
  </nz-radio-group>

  <div style="margin-top: 30px">
    <app-beijing *ngIf="radioValue == 'beijing'"></app-beijing>
    <app-shanghai *ngIf="radioValue == 'shanghai'"></app-shanghai>
    <app-guangzhou *ngIf="radioValue == 'guangzhou'"></app-guangzhou>
  </div>
</nz-card>

// app.component.ts
import { Component } from '@angular/core';

@Component({
  selector: 'app-root',
  templateUrl: './app.component.html',
  styleUrls: ['./app.component.css']
})
export class AppComponent {
  title = 'city';
  radioValue = 'beijing';
}
```

【代码解析】这个页面通过 nz-radio-group 创建了三个选项，通过参数 radioValue 和 ngIf 指令来控制显示与之对应的三个组件：app-beijing、app-shanghai、app-guangzhou。在把这三个组件都开发完毕后，读者试着单击一下，如果可以正常切换和显示组件，就可以进行下一步了。

（4）最后将三个组件开发完毕就大功告成了。代码如下：

```
// beijing.component.html
<h2>
  北京欢迎你！
</h2>
<p>
  北京，简称"京"，是中华人民共和国省级行政区、首都、直辖市，是全国的政治、文化中心和国际交往中心。北京地处中国华北地区，中心位于东经 116°20′、北纬 39°56′，东与天津毗连，其余均与河北相邻，北京市总面积 16410.54 平方千米。
</p>
<img style="width: 300px" src="./assets/beijing.png">

// shanghai.component.html
<h2>
  上海欢迎你！
</h2>
```

```
<p>
  上海，简称"沪"或"申"，是中国共产党的诞生地，是中华人民共和国省级行政区、直辖市，国家历史文化名城，国际经济、金融、贸易、航运、科技创新中心。 [1]  上海位于中国华东地区，界于东经120°52′-122°12′，北纬30°40′-31°53′之间，地处长江入海口，东隔东中国海与日本九州岛相望，南濒杭州湾，北、西与江苏、浙江两省相接，上海市总面积6340.5平方千米。
</p>
<img style="width: 300px" src="./assets/shanghai.png">

// guangzhou.component.html
<h2>
  广州欢迎你!
</h2>
<p>
  广州，简称穗，别称羊城、花城，是广东省省会、副省级市、国家中心城市、超大城市、国际大都市、国际商贸中心、国际综合交通枢纽、国家综合性门户城市，首批沿海开放城市，是南部战区司令部驻地[1-2]  。广州地处广东省中南部，珠江三角洲北缘，濒临南海，邻近香港、澳门，是中国通往世界的南大门，是粤港澳大湾区、泛珠江三角洲经济区的中心城市以及"一带一路"的枢纽城市。
</p>
<img style="width: 300px" src="./assets/guangzhou.png">
```

【代码解析】图片资源是笔者自己导入到项目中的，读者可以下载源代码，或者直接自行将三个城市的组件编辑成自己喜欢的样式，毕竟这是个练习，选择自己满意的样式即可。上面程序执行的结果如图 4.3~图 4.5 所示。

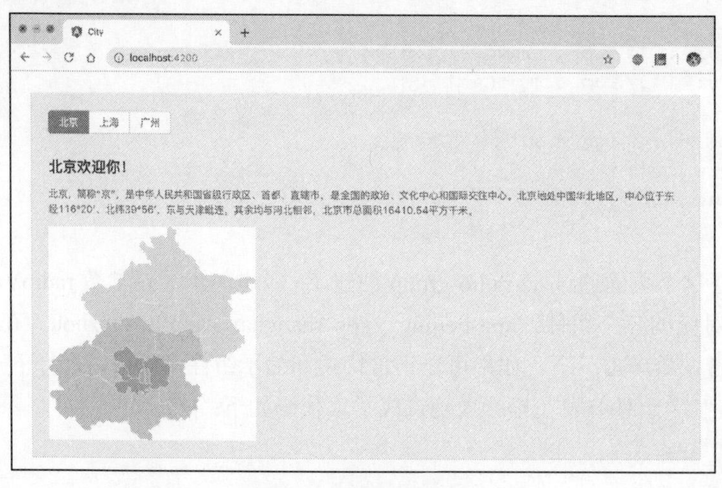

图 4.3　选择北京组件

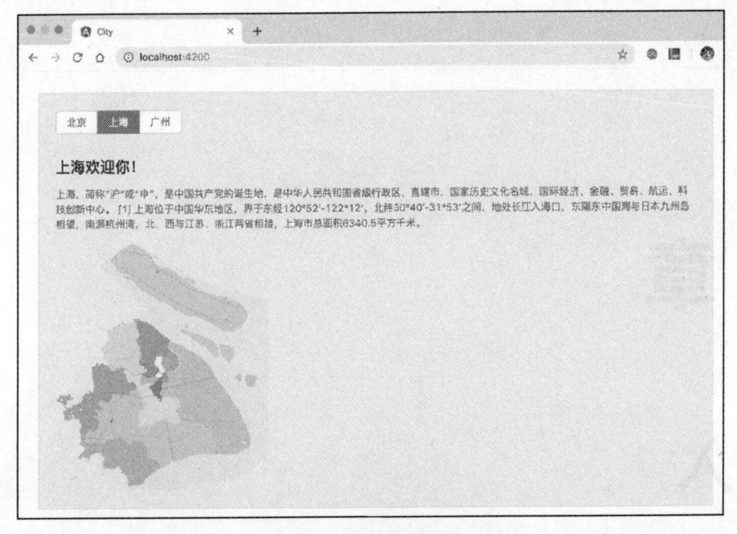

图 4.4　选择上海组件

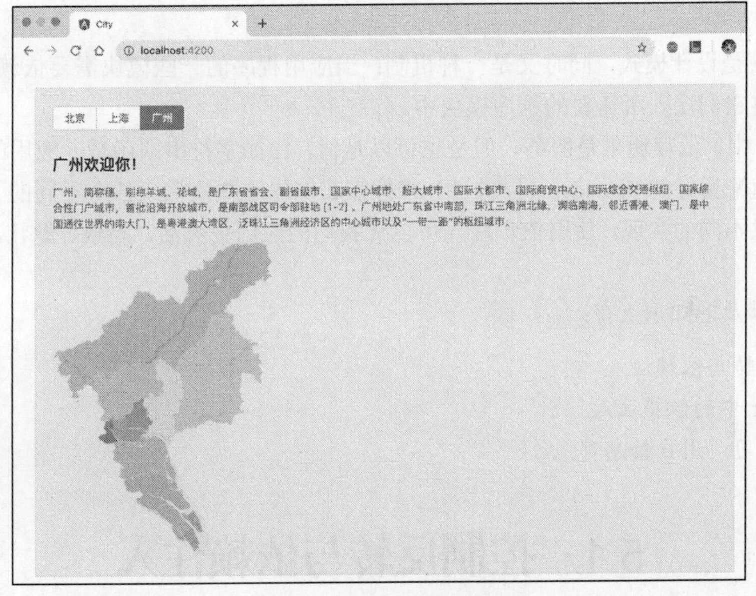

图 4.5　选择广州组件

4.6　小　结

本章的内容主要介绍了组件的组成、注解、生命周期，以及组件数据间的交互，加上最后的实战练习，相信读者已经基本掌握了组件的用法。组件是 Angular 中十分重要的一部分，说它难也不难，只要掌握了它的展示、传值后，其实就没有什么内容了，它实际上还是一种代码的封装和复用。如果暂时不理解也不需要担心，因为在接下来的内容中会经常使用组件，只要勤加练习就可以慢慢领会。

第5章

依赖注入

依赖注入既是设计模式，同时又是一种机制：当应用程序的一些模块需要依赖另一些模块时，利用依赖注入将它们注入所需要的程序模块中。

在 Angular 中，依赖通常是服务，但是也可以是值，比如字符串或函数。应用的注入器会使用为相应服务或值配置好的 Provider（提供者）来按需实例化这些依赖。各个不同的 Provider 可以为同一个服务提供不同的实现。使用依赖注入可以让我们的应用更灵活、高效、健壮，以及可测试和可维护。

本章主要涉及的知识点有：

- 控制反转与依赖注入
- Angular 中的依赖注入
- 实战练习：用户信息页

5.1 控制反转与依赖注入

本节首先从控制反转的基本概念开始介绍，再进行依赖注入的讲解，理解这些概念是学习使用 Angular 依赖注入的基础。了解了基本概念之后，我们才能从依赖注入的原理中找到学习的技巧。

5.1.1 控制反转

控制反转（Inversion of Control，缩写为 IoC）是面向对象编程的六大设计原则之一，主要用来降低代码之间的耦合度。我们通过下面的两张图来分析一下，其中图 5.1 是未使用控制反转的强耦合系统，图 5.2 是使用了控制反转后的系统。对比之下可以看出，控制反转这个“第三方”起了一个黏合剂的作用，使对象之间完成了解耦。

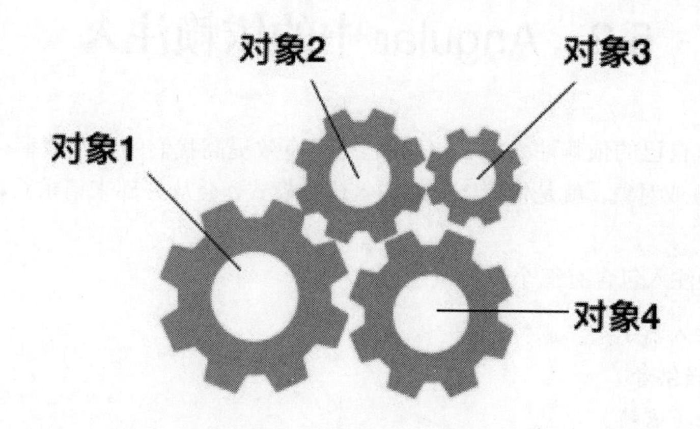

图 5.1　未使用控制反转

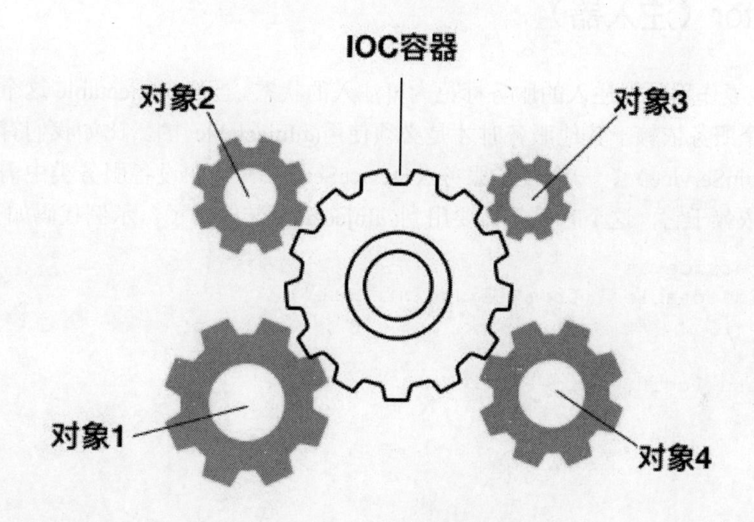

图 5.2　使用控制反转

刚才说的是概念上的，接下来就是实现了。控制反转常见的实现方式主要有两种，一种叫依赖查找（Dependency Lookup），另一种就是接下来要讲到的依赖注入（Dependency Injection，简称 DI）。

5.1.2　依赖注入

依赖注入是一种实现控制反转的手段，也是一种软件设计模式。在这种模式下，一个或更多的依赖（或服务）被注入（或者通过引用传递）到一个独立的对象（或客户端）中，然后成为该客户端状态的一部分。

5.2 Angular 中的依赖注入

在 Angular 中有自己的依赖注入框架，使用它可以有效提高我们的开发效率。在类执行功能时，所需要的外部的服务或对象，就是依赖。依赖注入这种模式，会从外部来请求获取，而不需要自己创建。

Angular 的依赖注入包含有三个重要概念：

- Injector（注入器）
- Provider（提供者）
- Dependence（依赖）

5.2.1 Injector（注入器）

Injector 的主要作用是把注入的服务标记为可注入的状态。当然@Injectable 这个装饰器不是必需的，只有在一个服务依赖于其他服务时才是必须使用@Injectable 的。比如我们有两个服务，一个登录服务（LoginService）、一个设备服务（DeviceService），在设备服务类中需要用到登录服务的一些功能而依赖于它，这个时候就需要用到@Injectable 装饰器了。示例代码如下：

```
// login.service.ts
import { Injectable } from '@angular/core';

@Injectable()
export class LoginService {

}



// device.service.ts
import { Injectable } from '@angular/core';
import { LoginService } from './login/login.service';

@Injectable()
export class DeviceService {
    constructor(private loginService: LoginService) {}
}

// app.component.ts
import { LoginService } from './login/login.service';
import { DeviceService } from './device/device.service';

providers: [
  DeviceService,
  LoginService
]
```

当然最好的做法是，无论是否依赖其他服务，都应该使用@Injectable 来装饰服务，这也是

Angular 官方所推荐的。

5.2.2 Provider（提供者）

Injector 通过 Provider 来创建被依赖对象的实例，没有在 Provider 中注入过的服务，是无法直接通过构造函数初始化来使用的。Angular 是一个层层嵌套的组件树，如图 5.3 所示。

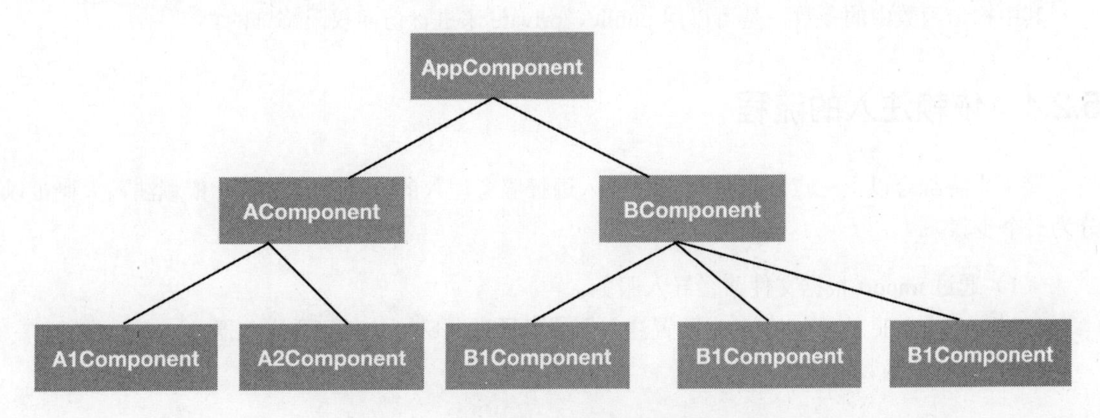

图 5.3 Angular 是一个层层嵌套的组件树

我们可以在任何一个层级注入，注入后该组件和它的子组件都会获得注入的实例，而不用重新注入。在很多 Angular 的入门教程中，都会提到要在 app.module 中写入服务，为什么要这么做呢？这里我们要知其所以然。在根组件 app.component 的 app.module 中注入服务后，就相当于在树的根部进行过了初始化，之后在每一个地方就都可以使用它了。

我们是应该把所有的组件都写在 app.module 中，还是写在使用的模块中呢？这是基于设计的原因，主要也是为了便于复用。虽然写在 app.module 中可以实现复用，但是这些服务都需要维持一个实例，会增加了一定的系统开销，因此最好的方法应该是：如果是单独使用的组件，就只在自己的模块中引入；如果是多个组件使用的，就在它们的父类组件中引入；如果是几乎在所有地方都会用到的组件，则在根模块中引入。

自从 Angular6 出现之后，对这部分进行了修改，除了之前的方式外。还可以在服务的 Injectable 中写入 providedIn 引入 root 即可完成全局设置，代码如下所示：

```
@Injectable({
  providedIn: 'root',
})
```

如果要引入指定的模块，就直接在 providedIn 写入模块名即可，代码如下所示：

```
@Injectable({
  providedIn: UserModule,
})
```

5.2.3 Dependence（依赖）

所谓的依赖，就是我们要用到的服务，在 Provider 中配置后，就可以在构造函数的声明中使用，如下所示。

```
constructor(public deviceService: DeviceService) {
}
```

其中构造函数中的参数，是可以用 public、private 来进行访问权限修饰的。

5.2.4 依赖注入的流程

接下来将综合以上三点，把使用依赖注入进行服务注入的流程进行讲解。依赖注入大概可以分为三个步骤。

（1）通过 import 根据文件路径导入服务。

（2）在 Module 的 providers 中配置注入器或者通过 providedIn 配置注入器。

（3）在构造函数中注入依赖。

5.3 实战练习：用户信息页

本节将构建一个工具类的服务，并结合上一章节学到的组件知识，开发一个用户信息验证的例子。由于现在还没有学到表单相关的知识，所以我们直接通过展示几个 input（输入），并使用依赖注入的服务进行验证。

（1）首先新建项目，并导入 NG-ZORRO。

```
ng new user-info
cd user-info
ng add ng-zorro-antd
```

（2）新建完成后，通过 Angular-CLI 新建一个用户组件和一个工具服务。由于新建的服务是不会自动导入的，所以新建后需要手动在 app.module.ts 中进行注册。

```
ng g component user-info
ng g service utils

// app.module.ts
import { BrowserModule } from '@angular/platform-browser';
import { NgModule } from '@angular/core';

import { AppComponent } from './app.component';
import { NgZorroAntdModule, NZ_I18N, zh_CN } from 'ng-zorro-antd';
import { FormsModule } from '@angular/forms';
import { HttpClientModule } from '@angular/common/http';
import { BrowserAnimationsModule } from
'@angular/platform-browser/animations';
import { registerLocaleData } from '@angular/common';
```

```
import zh from '@angular/common/locales/zh';
import { UserInfoComponent } from './user-info/user-info.component';
import { UtilsService } from './utils.service';

registerLocaleData(zh);

@NgModule({
  declarations: [
    AppComponent,
    UserInfoComponent
  ],
  imports: [
    BrowserModule,
    NgZorroAntdModule,
    FormsModule,
    HttpClientModule,
    BrowserAnimationsModule
  ],
  providers: [{ provide: NZ_I18N, useValue: zh_CN }, UtilsService],
  bootstrap: [AppComponent]
})
export class AppModule { }
```

（3）准备工作完成后，先来编写 utils.service.ts，这个文件主要用于存放一些工具类函数。代码如下：

```
import { Injectable } from '@angular/core';

@Injectable()
export class UtilsService {

  constructor() { }

  // 设置日期格式
  public formatDate(date: string) {
    const time = new Date(date);
    const datetime = time.getFullYear() + '-' +
this.formatDayAndMonth(time.getMonth() + 1) + '-' +
this.formatDayAndMonth(time.getDate());
    return datetime;
  }

  // 格式化日期和月份
  private formatDayAndMonth(val) {
    if (val < 10) {
      val = '0' + val;
    }
    return val;
  }

  // 手机号是否正确
  public isMobilePhone(value: string): boolean {
    let reg: RegExp = /^1[3|4|5|8][0-9]\d{4,8}$/;
    return this.regVerify(value, reg);
  }
```

```
    // 邮箱是否正确
    public isEmail(value: string): boolean {
      let req: RegExp =
/^[a-z0-9]+([._\\-]*[a-z0-9])*@([a-z0-9]+[-a-z0-9]*[a-z0-9]+.){1,63}[a-z0-9]+$
/;
      return this.regVerify(value, reg);
    }

    // 正则表达式验证
    public regVerify(value: string, reg: RegExp): boolean {
      if (reg.test(value)) {
        return true;
      } else {
        return false;
      }
    }
  }
```

【代码解析】注释已经明确说明了各个方法的职责。其中手机号和邮箱验证都使用了下面的 regVerify 方法，新增的方法一是为了封装，二是为了给外部提供一个可以手动控制正则表达式的方式。

（4）接下来编写 user-info.component.html 的内容。代码如下：

```
  <input nz-input placeholder="手机号" [(ngModel)]="phoneNumber">
  <button (click)="verifyPhoneNumber()" class="top" nz-button nzType="primary">
验证手机号</button>

  <input class="top" nz-input placeholder="电子邮箱" [(ngModel)]="email">
  <button (click)="verifyEmail()" class="top" nz-button nzType="primary">验证
电子邮箱</button>

  <nz-date-picker class="top" [(ngModel)]="date"
(ngModelChange)="onChange($event)"></nz-date-picker>
  <div class="top">生日：{{birthday}}</div>
  <button (click)="changeDate()" class="top" nz-button nzType="primary">转换日
期</button>
```

【代码解析】这里展示了三个不同的例子，手机号验证、邮箱验证和日期格式化。按钮的单击事件将在 TS 代码中完成。

（5）最后将 user-info.component.ts 中的实现代码编写完毕，就可以看到结果了。代码如下：

```
  import { Component, OnInit } from '@angular/core';
  import { UtilsService } from '../utils.service';

  @Component({
    selector: 'app-user-info',
    templateUrl: './user-info.component.html',
    styleUrls: ['./user-info.component.css']
  })
  export class UserInfoComponent implements OnInit {
    // 手机号
    phoneNumber: string = '';
```

```
  // 电子邮箱
  email: string = '';
  // 日期选择
  date: string = '';
  // 生日
  birthday: string = '';

  constructor(private util: UtilsService) { }
  ngOnInit() {
  }

  // 验证手机号
  verifyPhoneNumber() {
    if (this.util.isMobilePhone(this.phoneNumber)) {
      console.log('这是正确的手机号');
    } else {
      console.log('这不是正确的手机号');
    }
  }

  // 验证电子邮箱
  verifyEmail() {
    if (this.util.isEmail(this.email)) {
      console.log('这是正确的电子邮箱');
    } else {
      console.log('这不是正确的电子邮箱');
    }
  }

  // 选择日期
  onChange() {
    this.birthday = this.date;
  }

  // 格式化日期
  changeDate() {
    this.birthday = this.util.formatDate(this.date);
  }
}
```

【代码解析】头部导入 UtilsService，通过最基本的依赖注入构造函数中。之后通过按钮的单击事件 verifyPhoneNumber、verifyEmail 来实现验证，结果会在控制台输出。最后的日期会在选择后先显示在生日栏，单击格式化后会直接修改结果。程序的执行结果如图 5.4 和图 5.5 所示。

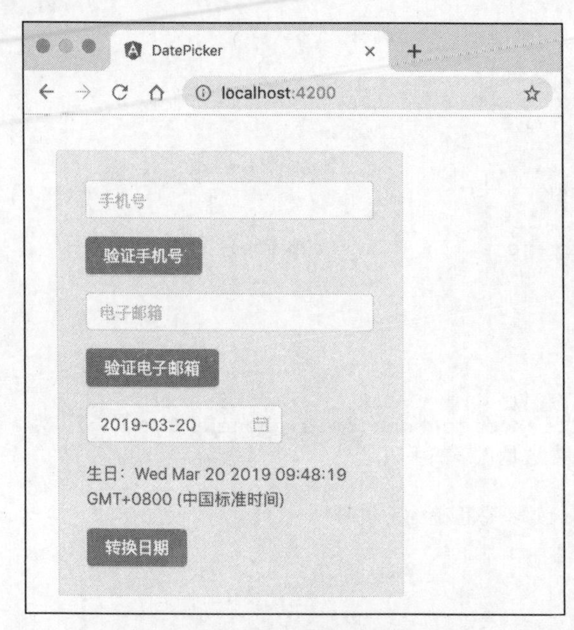

图 5.4　未转换的时间

图 5.5　转换后的时间

5.4 小　结

本章中从面向对象编程的设计原则之一“控制反转”入手，结合设计模式中依赖注入，对Angular 中依赖注入的概念及用法进行了介绍，最后通过一个实战例子将所学内容进行了复习和巩固。本章的内容不是很多，相信读者通过对本章的学习，对 Injector（注入器）、Provider（提供者）有一定的了解，并掌握了依赖注入的基本用法。下一章将对 HTTP 网络请求部分进行讲解。

第 6 章

HTTP

HTTP（HyperText Transfer Protocol）通常被称为超文本传输协议，又译为超文本转移协议。HTTP 作为应用最为广泛的网络协议，不论前端开发者和后端开发者都会涉及，在 Angular 中当然也需要发送网络请求，所以本章的知识点也是必须掌握的。

本章主要涉及的知识点有：

- 如何发送网络请求
- HTTP 协议基础知识
- HTTP 与 HTTPS 的区别
- Angular 中 HTTP 的改动
- 实战练习：制作一个 HTTP 拦截器

6.1 HTTPClient——发送第一条网络请求

如本章开头所述，HTTP 是前端开发者经常会接触到的协议，我们通过它与远程服务端进行通讯交互，而在 Angular 中需要使用 HttpClient 来实现这个功能。使用 HttpClient 发送的所有请求，都会返回一个 Observable，所以我们请求时必须调用 subscribe。本节首先用尽量少的代码，指导读者先成功地发送一条网络请求，先学会如何使用，再详细讲解其中的原理。

在 Angular 中发送网络请求，需要使用它自带的 HttpClient。下面输入以下指令新建一个项目。

```
ng new http
```

【示例 6-1】发送一条 GET 请求。项目创建完成后，等待 npm 安装依赖，之后在 app.module.ts 导入 HttpClient。

```
import { BrowserModule } from '@angular/platform-browser';
import { NgModule } from '@angular/core';
```

```
import { AppComponent } from './app.component';
import { HttpClientModule } from '@angular/common/http';

@NgModule({
  declarations: [
    AppComponent
  ],
  imports: [
    BrowserModule,
    HttpClientModule
  ],
  providers: [],
  bootstrap: [AppComponent]
})
export class AppModule { }
```

在 app.component.ts 输入以下代码，使用 http://jsonplaceholder.typicode.com/ 这个网站提供的测试接口进行网络请求。

```
import { Component } from '@angular/core';
import { HttpClient } from '@angular/common/http';

@Component({
  selector: 'app-root',
  templateUrl: './app.component.html',
  styleUrls: ['./app.component.css']
})
export class AppComponent {
  title = 'http';

  constructor(private http: HttpClient) {}

  ngOnInit() {
    this.http.get('https://jsonplaceholder.typicode.com/todos/1').subscribe(
(res) => {
      console.log(res);
    });
  }
}
```

【代码解析】导入 Angular 提供的 HttpClient，并通过构造函数注入。在 ngOnInit 中发送最基本的 GET 请求。打开 Chrome 的控制台，在 Network 栏可以看到网络请求已经成功发送了，如图 6.1 所示。

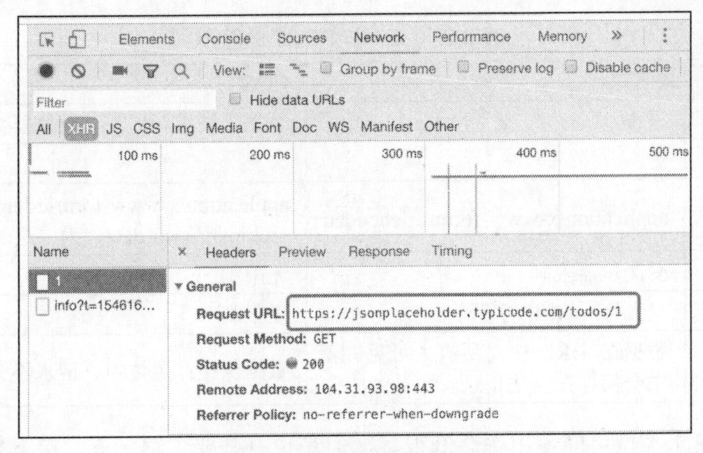

图 6.1 发送 GET 网络请求

6.2 HTTP 协议基础知识

在进行前端开发时，我们一般需要完成和后台进行接口对接的工作，不管服务端是使用 Java、PHP、Python 或是其他语言编写的，都是通过 HTTP 协议来进行交互的。我们每对 HTTP 协议多一分理解，就能少一分前端开发者和后端开发者在交流上的障碍。

6.2.1 请求方法

首先通过表格列举一下不同请求方法的应用场景，参见表 6.1。

表 6.1 常见 HTTP 请求方法的应用场景

方法	说明
GET	获取资源
POST	传输实体主体
PUT	传输替换目标资源
DELETE	删除目标资源
HEAD	获取报文首部
OPTIONS	获取支持的方法

很多小公司的服务端开发，一般不怎么遵守 RESTful API 的设计，很可能所有接口都收 POST，或者说只有 GET 和 POST，这种做法在使用上是没有影响的，只是不规范。这两种请求类型也是最常见的，所以先来讲一下 GET 和 POST 这两个请求方法的区别，下面直接整理成一个表格，方便读者查看，如表 6.2 所示。

表 6.2　GET 与 POST 的区别

	GET	POST
后退按钮/刷新	无害	数据会被重新提交（浏览器会提示）
书签/缓存/历史	√	×
编码类型	application/x-www-form-urlencoded	application/x-www-form-urlencoded 或 multipart/form-data。为二进制数据使用多重编码
数据长度限制	受浏览器限制	无限制
数据类型限制	只允许 ASCII 字符	无限制
可见性	数据在 URL 中对所有人可见、请求会保存在历史记录	数据保存在主体中、请求不会保存在历史记录

虽然其他的请求类型可能使用率会低很多，但是也有必要了解一下。接下来我们对其他的请求方法进行简单的说明。

（1）PUT 请求

PUT 与 POST 方法的区别在于，PUT 方法是幂等的：调用一次与连续调用多次是等价的（即没有副作用），而连续调用多次 POST 方法可能会有副作用，比如将一个订单重复提交多次。

（2）DELETE 请求

如果 DELETE 方法成功执行，那么可能会有以下几种状态码：

- 状态码 202(Accepted) 表示请求的操作可能会成功执行，但是尚未开始执行。
- 状态码 204(No Content)表示操作已执行，但是无进一步的相关信息。
- 状态码 200(OK) 表示操作已执行，并且在响应中提供了相关状态的描述信息。

（3）HEAD 请求

HEAD 请求就如同前面表格所说的是获取报文首部，该请求方法的一个使用场景是在下载一个大文件前先获取其大小再决定是否要下载，以此节约网络带宽资源。

（4）OPTIONS 请求

OPTIONS 请求就是用于获取目的资源所支持的通信选项，平时在执行 Ionic、Angular 的时候，它就经常会在请求前自动调用一个 OPTIONS 方法。

6.2.2　HTTP 状态码

简单地说，HTTP 状态码就是描述返回的请求结果。下面还是用一个表格来展示一下，由于种类比较多，所以只列举种类，不进行完全列举，如表 6.3 所示。

表 6.3　HTTP 状态码

	类别	原因
1xx	Informational（信息性状态码）	接收的请求正在处理
2xx	Success（成功状态码）	请求正常处理完毕
3xx	Redirection（重定向状态码）	需要进行附加操作以完成请求

（续表）

	类别	原因
4xx	Client Error（客户端错误状态码）	服务器无法处理请求
5xx	Server Error（服务端错误状态码）	服务器处理请求出错

在开发过程中，2xx、4xx 和 5xx 应该更常见一些，知道这些状态码的含义，能更好地分析出现问题的原因。

6.2.3 请求报文首部

【示例 6-2】分析 HTTP 的请求报文首部。HTTP 报文的首部字段，主要是用来传递额外的重要信息，我们举一个简单的例子。

```
// 发起请求
GET / HTTP/1.1
Request URL: https://www.baidu.com/favicon.ico
Host: www.baidu.com
Accept-Language: zh-CN

// 服务端返回
HTTP/1.1 200 OK
Date: Sat, 07 Apr 2018 02:17:48 GMT
Server: Apache
Last-Modified: Mon, 02 Apr 2018 09:39:34 GMT
Accept-Ranges: bytes
Content-Length: 984
Content-Type: image/x-icon
```

这些参数都是用来传递额外信息的，我们再带上注释解释一下这些信息。

```
// 发起请求
// 请求方法 / HTTP 版本号
GET / HTTP/1.1
// 请求地址
Request URL: https://www.baidu.com/favicon.ico
// 请求资源所在服务器
Host: www.baidu.com
// 优先选择的语言（自然语言）
Accept-Language: zh-CN

// 服务端返回
// HTTP 版本、HTTP 状态码 200
HTTP/1.1 200 OK
// 创建报文的日期
Date: Sat, 07 Apr 2018 02:17:48 GMT
// HTTP 服务器的安装信息
Server: Apache
// 资源的最后修改时间
Last-Modified: Mon, 02 Apr 2018 09:39:34 GMT
// 支持字节范围请求
Accept-Ranges: bytes
```

```
// 实体主体的大小
Content-Length: 984
// 实体主体的类型
Content-Type: image/x-icon
```

HTTP 首部字段种类非常多，该例子列举了常用的一部分，若想了解更多种类，可以查看 MDN 上的 HTTP Headers 文档，网址为 https://developer.mozilla.org/zh-CN/docs/Web/ HTTP/Headers，如图 6.2 所示。

HTTP Headers

跳转到： Notes See also

Web 技术文档 › HTTP › HTTP Headers

这篇翻译不完整。请帮忙从英语翻译这篇文章。

图 6.2 MDN HTTP Headers

从图 6.2 中可以看出，MDN 里的一部分内容中文翻译还不完整，翻译水平高的朋友可以帮忙完善一下，为开源事业出一份力吧。

讲解完 HTTP 的请求报文首部，我们还需要知道如何在 Angular 中去设置它们。接着上文的例子进行开发，输入以下代码：

```
import { Component } from '@angular/core';
import { HttpClient, HttpHeaders } from '@angular/common/http';

@Component({
  selector: 'app-root',
  templateUrl: './app.component.html',
  styleUrls: ['./app.component.css']
})
export class AppComponent {
  title = 'http';

  constructor(private http: HttpClient) {}

  ngOnInit() {
    let headers = new HttpHeaders();
    headers = headers.set('Content-Type', 'application/json; charset=utf-8');
    this.http.get('https://jsonplaceholder.typicode.com/todos/1', { headers: 
headers }).subscribe((res) => {
      console.log(res);
    });
  }
}
```

【代码解析】改动其实不大，主要就是通过 import 导入了 HttpHeaders，并在请求之前创建一个 headers 对象，再通过 SET 方法进行了设置。这里举的例子是 Content-Type，其实其他的请求报文首部都是一样的，通过“键-值对”（Key-Value Pair）进行设置即可。

6.3 HTTP与HTTPS

HTTP与HTTPS经常使人混淆，本节详细介绍HTTPS，让读者找到两者的区别。

6.3.1 为什么需要HTTPS

在安全方面，HTTP存在以下三个方面的缺点。

- **窃听风险：**由于通信使用明文传输，内容可能会泄露。
- **篡改风险：**第三方对传输的数据进行篡改，影响与服务端之间的正确通信。
- **冒充风险：**可能会出现中间人攻击、第三方冒充服务器等情况。

其实上面三个方面的问题可以总结为通信未加密，而HTTPS的出现则很好地解决了以上问题。

6.3.2 什么是 HTTPS

与HTTP协议的明文传输相比，HTTPS是将这些内容进行加密，确保信息传输的安全。最后一个字母S指的是SSL（Secure Socket Layer，安全套接层）/ TLS（Transport Layer Security，安全传输层协议）协议，它位于HTTP协议与TCP/IP协议之间。

HTTPS使用了非对称加密（就是有一对公钥和私钥）。服务器的私钥只存在于服务器上，服务器下发的内容不可能被伪造，因为别人都没有私钥，所以无法进行数字签名。所有人都有自己的公钥，但服务器用于数字签名的私钥只有服务器有，所以服务器才能对要发送的内容用自己的私钥进行数字签名。

HTTPS还可以对要发送的内容进行加密，就是用接受方的公钥进行加密再发送给接收方，以确保信息传输的安全，接收方接到加密的内容后可以用自己的私钥进行解密，其他人获得被加密的内容是无法解密的，因为只有接收方拥有自己的私钥，只有用自己的私钥才可以解密自己公钥加密的内容。不过，在实际应用中，不会用非对称加密法直接加密要传输的内容，因为计算量太大了，如果传输的内容多了，计算量会成为负担，所以要传输的内容一般会用对称加密算法进行加密，而使用非对称加密方法来传输对称加密算法中用到的密钥（或称为密码）。现在很多网络应用的传输都在逐步转为HTTPS。

6.3.3 HTTPS工作过程

针对6.3.2节提到的三个方面的缺点，本节说明一下HTTPS是如何应对的。

HTTPS使用非对称加密算法来传输对称加密算法用的密码，而传输的内容则使用这个对称加密算法的密码进行加密，用对称加密和非对称加密相结合的方法来避免第三方获取内容，做到了既保证了加密的强度又兼顾了加密的效率。

- 发送方将要发送内容的哈希值用自己的私钥进行加密（即提供自己的数字签名），再和要发送的内容一起发送给接收方，接收方会把收到的数字签名用发送方的公钥进行解密，得到还原后的哈希值，再和自己收到的内容计算出的新哈希值进行对比，如果这两组哈希值相同，就表示收到的内容没有被篡改过，否则就是收到的内容中途被人篡改了。这个基于严谨的数学逻辑的加密解密过程就可以避免传送的内容在传输途中被人篡改而一无所知。
- HTTPS 需要使用由权威机构发布的 CA（Certificate Authority，电子商务认证授权机构）证书，并通过证书校验机制防止第三方冒名顶替。

提 示

哈希值是将内容（数据或信息）通过哈希算法提取后得到的数据值，理论上来说不管多复杂的内容都可以通过哈希算法求得哈希值。比如下载的 Android SDK 就会提供一个 SHA-256 哈希值，它是哈希算法的一种，如图 6.3 所示。

平台	Android Studio 软件包	大小	SHA-256 校验和
Windows（64位）	android-studio-ide-173.4720617-windows.exe Recommended	758 MB	e2695b73300ec398325cc5f242c6ecfd6e84db190b7d48e6e78a8b0115d49b0d
	android-studio-ide-173.4720617-windows.zip No .exe installer	854 MB	e8903b443dd73ec120c5a967b2c7d9db82d8ffb4735a39d3b979d22c61e882ad
Windows（32位）	android-studio-ide-173.4720617-windows32.zip No .exe installer	854 MB	c238f54f795db03f9d4a4077464bd9303113504327d5878b27c9e965676c6473
Mac	android-studio-ide-173.4720617-mac.dmg	848 MB	4665cb18c838a3695a417cebc7751cbe658a297a9d6c01cbd9e9a1979b8b167e
Linux	android-studio-ide-173.4720617-linux.zip	853 MB	13f290279790df570bb6592f72a979a495f7591960a378abea7876ece7252ec1

图 6.3　SHA-256

6.3.4　申请 HTTPS

现在有很多网站都已经广泛使用 HTTPS，比如 www.baidu.com，用 Chrome 就能看到地址栏中的 HTTPS 证书以及安全的标识，如图 6.4 所示。

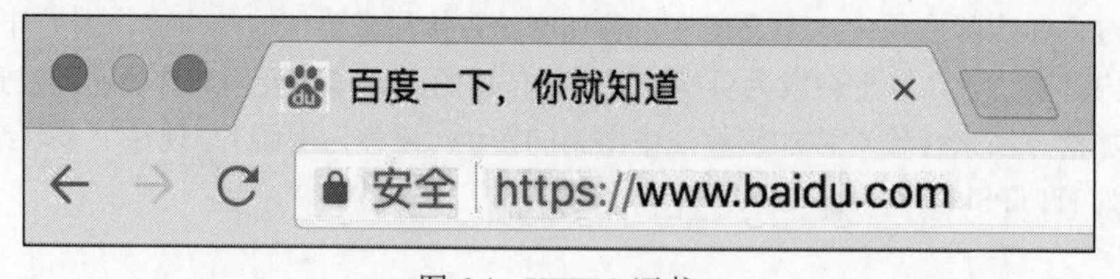

图 6.4　HTTPS 证书

而且现在 iOS 提交至 App Store 的应用都必须使用 HTTPS 进行网络请求。因此了解如何使用 HTTPS 还是很有必要的。

申请 https 证书的方式很简单，找到卖 HTTPS 证书的网站，找到 CA 证书服务，填写信息购买即可，如图 6.5 所示。

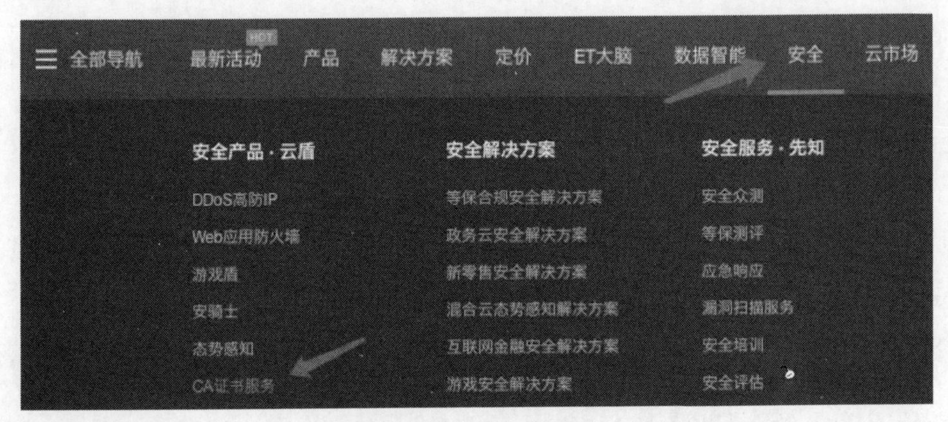

图 6.5 购买 HTTPS 证书

6.3.5 为什么不一直使用 HTTPS

使用 HTTP 协议的交互过程有“三次握手”（Three Times Handshake），即在建立收发两端的网络连接前的协商过程。在加入 HTTPS 之后就变成了四次握手，即 http 上增加了 SSL，所以效率会降低一些，不过还是能接受的。

由于 HTTPS 会降低一定的速度，还有一些额外的成本，因此对于一些不太需要加密的信息，还是有很多企业倾向于选择 HTTP。

6.4 实战练习：制作一个 HTTP 拦截器

HTTP 拦截器在开发过程中十分常见。在构建项目架构时，最好就创建好 HTTP 拦截器，否则遇到以下几种问题，再进行改动就会十分浪费时间。

需要给所有的请求修改请求地址。

需要给所有请求参数设置新的请求报文首部。

需要监听所有请求的状态码。

（1）输入以下指令新建一个项目。

```
ng new interceptor
```

（2）项目创建完成后，等待 npm 安装依赖，之后在 app.module.ts 导入 HttpClient。

```
import { BrowserModule } from '@angular/platform-browser';
import { NgModule } from '@angular/core';

import { AppComponent } from './app.component';
import { HttpClientModule } from '@angular/common/http';

@NgModule({
```

```
  declarations: [
    AppComponent
  ],
  imports: [
    BrowserModule,
    HttpClientModule
  ],
  providers: [],
  bootstrap: [AppComponent]
})
export class AppModule { }
```

（3）新建一个文件 http-service.ts 并输入以下代码：

```
import { HttpClient } from '@angular/common/http';
import { Injectable } from '@angular/core';
import { Observable, of } from 'rxjs';
import { catchError, map, tap, retry } from 'rxjs/operators';
@Injectable()
export class HttpService {
  private baseUrl = 'https://jsonplaceholder.typicode.com/';

  constructor(public http: HttpClient) { }
  private extractData(res: Response) {
    const body = res;
    return body || {};
  }

  private handleError<T>(operation = 'operation', result?: T) {
    return (error: any): Observable<T> => {
      console.error(error); // log to console instead
      console.log(`${operation} failed: ${error.message}`);
      return of(result as T);
    };
  }
}
```

【代码解析】 baseUrl 是网络请求通用的地址，这样发送网络请求时就不需要重复写了。extractData 这个方法是用来对请求参数进行处理的，在这个例子中，我们虽然不需要处理请求参数，但是我们还是把这部分预留出来了。handleError 用来截获异常情况。这几个方法不需要对外公布，所以使用 private 进行修饰。

（4）之后写上需要用到的请求方法，比如 POST 和 GET。http.service.ts 最终的代码如下：

```
import { HttpClient } from '@angular/common/http';
import { Injectable } from '@angular/core';
import { Observable, of } from 'rxjs';
import { catchError, map, tap, retry } from 'rxjs/operators';
@Injectable()
export class HttpService {
  private baseUrl = 'https://jsonplaceholder.typicode.com/';
```

```
  constructor(public http: HttpClient) { }
  public get(url: string): Observable<any> {
    return this.http.get(this.baseUrl + url).pipe(
      retry(2),
      map(this.extractData),
      tap(() => { }),
      catchError(this.handleError('get', []))
    );
  }

  public post(url: string, data: any = {}): Observable<any> {
    return this.http.post(this.baseUrl + url, data).pipe(
      retry(2),
      map(this.extractData),
      tap(() => { }),
      catchError(this.handleError('post', []))
    );
  }

  private extractData(res: Response) {
    const body = res;
    return body || {};
  }
  private handleError<T>(operation = 'operation', result?: T) {
    return (error: any): Observable<T> => {
      console.error(error); // log to console instead
      console.log(`${operation} failed: ${error.message}`);
      return of(result as T);
    };
  }

}
```

【代码解析】在 GET 和 POST 请求中都是将 baseUrl 拼接在前面，retry 是网络请求失败时重试的次数，map 中存放了我们的请求参数。这个网络请求最终会返回一个可观察对象 Observable。

（5）最后回到 app.component.html 并输入以下代码，进行网络请求测试。

```
import { Component } from '@angular/core';
import { HttpService } from './http-service';

@Component({
  selector: 'app-root',
  templateUrl: './app.component.html',
  styleUrls: ['./app.component.css']
})
export class AppComponent {
  title = 'interceptor';

  constructor(private http: HttpService) {
    this.http.get('todos/1').subscribe(res => {
      console.log(res);
    });
  }
}
```

（6）打开控制台查看 Console（控制台），我们可以看到 url 已经顺利拼接在前面，并成功地输出了返回值，如图 6.6 所示。

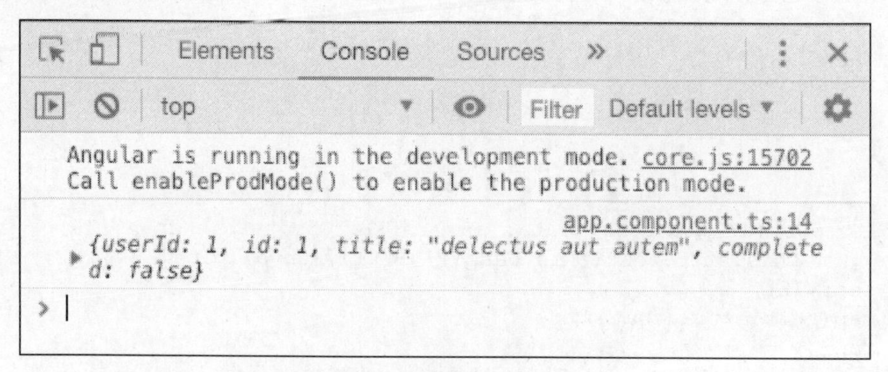

图 6.6 使用拦截器发送网络请求

6.5 小 结

本章首先带领读者学习如何简单快速地发起一个网络请求，之后对 HTTP 协议进行了一番讲解，并指出如何在 Angular 中使用它们。最后的实战练习开发了一个 HTTP 拦截器。相信读者已经掌握了如何发送网络请求。下一章我们将对开发中十分常见的表单部分进行讲解。

第7章 表单

商业项目总离不开数据，而数据会依赖于不同的表现形式，不管是图片、表格，还是我们常见的文本框、时间选择器，都是通过不同的表现形式提供给网络浏览者，让他们对数据进行阅读和分析，因此表单的设计就是一种数据的展示或表现形式，由此可见我们如何构建让用户阅读顺畅、操作方便的表单就显得十分重要。

本章主要涉及的知识点有：

- Angular 中的表单
- 实战练习：响应式表单
- 实战练习：模板驱动型表单

7.1 Angular 中的表单

在我们所创建的前端项目中，使用表单处理用户的输入是十分常见的基本功能。无论从用户的注册、登录，还是复杂的表单提交，都需要对所要提交的数据进行验证，才能让程序的运行流程更加完善。本章需要讲解的内容比较少，设计表单处理用实际的例子去上手实战练习，学习的效果才更好。

7.1.1 响应式表单与模板驱动型表单

Angular 提供了两种不同的方法来通过表单处理用户输入，分别是响应式表单和模板驱动型表单。它们的区别参见表 7.1。

表 7.1 响应式表单与模板驱动型表单的区别

	响应式表单	模板驱动型表单
表单模式	显式，在组件类中创建	隐式，由组件创建
数据模式	结构化	非结构化
可预测性	同步	异步
可扩展性	访问底层 API	在 API 之上的抽象
表单验证	函数	指令
是否可变	不可变	可变

相比之下，响应式表单通过提供对底层 API 的访问和对表单模型的同步访问，让创建大型表单变得更为轻松。模板驱动型表单则更加专注于简单的场景，它不可重用、对底层 API 进行抽象，而且对表单模型的访问是异步的。至于使用哪一种来进行表单的构建，就要看具体的需求场景了。

7.1.2 FormBuilder

响应式表单和模板驱动型表单共享了一些底层构造块。

- FormControl：用于追踪单个表单控件的值和验证状态。
- FormGroup：用于追踪一个表单控件组的值和状态。
- FormArray：用于追踪表单控件数组的值和状态。
- ControlValueAccessor：用于在 Angular 的 FormControl 实例和原生 DOM 元素之间创建一个桥梁。

表单是一个很重要的内容，概念性的东西就不多讲了。在下节我们做一次实战练习，来巩固对表单知识的学习。

7.2 实战练习：模板驱动型表单

表单这个部分的内容比较偏向实践，本节我们直接进行实战练习，首先从较为简单的模板驱动型表单开始。

7.2.1 建立模板驱动型表单项目

输入以下指令新建项目，并导入 ng-zorro。

```
ng new form
cd form
ng add ng-zorro-antd
```

因为模板驱动型表单更加适用于较为简单的场景，所以接下来以最常见的登录表单作为例子来进行讲解。输入以下代码新建组件，并在 app.module.ts 导入 FormsModule。

```
ng g component login --spec=false

// app.module.ts
import { BrowserModule } from '@angular/platform-browser';
import { NgModule } from '@angular/core';

import { AppComponent } from './app.component';
import { NgZorroAntdModule, NZ_I18N, zh_CN } from 'ng-zorro-antd';
import { FormsModule } from '@angular/forms';
import { HttpClientModule } from '@angular/common/http';
import { NoopAnimationsModule } from '@angular/platform-browser/animations';
import { registerLocaleData } from '@angular/common';
import zh from '@angular/common/locales/zh';
import { LoginComponent } from './login/login.component';

registerLocaleData(zh);

@NgModule({
  declarations: [
    AppComponent,
    LoginComponent
  ],
  imports: [
    BrowserModule,
    NgZorroAntdModule,
    FormsModule,
    HttpClientModule,
    NoopAnimationsModule
  ],
  providers: [{ provide: NZ_I18N, useValue: zh_CN }],
  bootstrap: [AppComponent]
})
export class AppModule { }
```

最后将 app.component.html 的代码连接到 login 组件。

```
<app-login></app-login>
```

7.2.2　在登录组件实现模板驱动型表单

准备工作就绪了，接下来在 login.html 和 login.css 输入以下代码构建页面。

```
// login.html
<form (ngSubmit)="onSubmit()" #loginForm="ngForm">
  <div style="width: 300px;margin: 50px 50px">
    <input class="distance" type="text" nz-input placeholder="请输入账号"
required [(ngModel)]="username" name="name"
      #name="ngModel">
    <input class="distance" type="password" nz-input placeholder="请输入密码"
required [(ngModel)]="passowrd"
      name="pwd" #pwd="ngModel">
    <label class="distance" nz-checkbox [(ngModel)]="remember" name="check">
      <span>记住我</span>
    </label>
```

```
    <a class="distance" style="float:right">忘记密码</a>
    <button type="submit" [disabled]="!loginForm.form.valid" class="distance"
nz-button [nzType]="'primary'" style="width: 300px">登录</button>
  </div>
</form>

// login.css
.distance {
  margin-top: 20px;
}
```

【代码解析】首先用 form 标签包裹所有与表单相关的内容，并设置提交方法与表单名。与之相呼应的是下面的登录按钮，它把 type 设置为 submit 即可响应 form 标签的提交方法，然后在按钮上设置了 disabled 为“!loginForm.form.valid”，这句话的作用是 loginForm 的表单未被验证通过时，让该按钮不可用。

我们分别把账号字段、密码字段设置为必填项，“记住密码”选项设置为非必选项。但是有一点需要注意，这三项除了设置 ngModel 外，还给它们设置了 name 属性。因为在 Angular 表单中使用 ngModel 的时候，不给它加一个 tag 是会报错的，如图 7.1 所示。

```
▸ERROR Error: If ngModel is used within a form tag, either the name attribute must be set or    LoginComponent.html:3
the form
      control must be defined as 'standalone' in ngModelOptions.

      Example 1: <input [(ngModel)]="person.firstName" name="first">
      Example 2: <input [(ngModel)]="person.firstName" [ngModelOptions]="{standalone: true}">
    at Function.push../node_modules/@angular/forms/fesm5/forms.js.TemplateDrivenErrors.missingNameException (forms.js:
4385)
```

图 7.1　ngModel 必须设置 tag

最后在 login.ts 中写入以下代码：

```
import { Component, OnInit } from '@angular/core';

@Component({
  selector: 'app-login',
  templateUrl: './login.component.html',
  styleUrls: ['./login.component.css']
})
export class LoginComponent implements OnInit {

  username: string = '';
  passowrd: string = '';
  remember: boolean = false;

  constructor() { }

  ngOnInit() {
  }

  onSubmit() {
    alert(`账号：${this.username} 密码：${this.passowrd} 记住密码：
${this.remember}`);
  }
}
```

【代码解析】新建了三个属性，并在 onSubmit 方法中输出了它们的值。未输入账号、密码之前，“登录”按钮处于不可用状态，输入后方可用。这个程序的运行过程如图 7.2 和图 7.3 所示。

图 7.2　未填写账号、密码时，“登录”按钮处于不可用状态

图 7.3　已填写账号、密码，“登录”按钮变为可用状态

单击“登录”按钮，弹出的对话框如图 7.4 所示。

图 7.4　单击“登录”按钮之后，弹出的对话框

除了设置必填项以外，还有其他的验证方式，比如最小长度、最大长度等，修改 login.html，增加最小长度的设置，并增加动态提示。

```
    <form (ngSubmit)="onSubmit()" #loginForm="ngForm">
      <div style="width: 300px;margin: 50px 50px">
        <input class="distance" type="text" nz-input placeholder="请输入账号"
required minlength="6" [(ngModel)]="username" name="name"
          #name="ngModel">
        <div *ngIf="name.errors?.minlength">
          账号必须大于 6 位
        </div>
        <input class="distance" type="password" nz-input placeholder="请输入密码"
```

```
required minlength="6" [(ngModel)]="passowrd"
          name="pwd" #pwd="ngModel">
        <div *ngIf="pwd.errors?.minlength">
          密码必须大于 6 位
        </div>
        <label class="distance" nz-checkbox [(ngModel)]="remember" name="check">
          <span>记住我</span>
        </label>
        <a class="distance" style="float:right">忘记密码</a>
        <button type="submit" [disabled]="!loginForm.form.valid" class="distance"
nz-button [nzType]="'primary'" style="width: 300px">登录</button>
      </div>
    </form>
```

【代码解析】在需要设置最小长度的 input 标签中设置 minlength 属性即可。之后在其标签底部写上提示信息，并使用 ngIf 控制其显示。name.errors?.minlength 所显示的条件就是不满足最小长度，问号是 Angular 的特殊语法，没有值的时候不提取值以防止报错，并且在内容为空时不进行提示。最终运行结果如图 7.5 所示。

图 7.5　设置 input 最小长度

因为 name.errors?.minlength 是动态的，所以只要内容长度大于 6，就会立刻隐藏。

7.3　实战练习：响应式表单

讲完了模板驱动型表单，接下来再做一个响应式表单的实战练习。响应式表单的内容更加丰富、扩展性也更强，所以在日常开发中也更常见。

7.3.1　创建响应式表单项目

继续在 form 项目中进行开发，输入以下指令新建一个组件来编写响应式表单的内容。

```
ng g component form-test --spec=false
```

之后在 app.component.html 中加入 form-test 组件，之前的登录组件就先注释掉。

```
<!-- <app-login></app-login> -->

<nz-card style="margin: 30px; width: 500px; background-color: #f2f4f5">
  <app-form-test></app-form-test>
</nz-card>
```

7.3.2 使用响应式表单构建个人资料页

首先编写 form-test.component.html 的代码，将表单页面构建完成。

```
<form nz-form [formGroup]="validateForm" (ngSubmit)="submitForm()">
  <nz-form-item>
    <nz-form-label [nzSm]="6" [nzXs]="24" nzFor="nickname" nzRequired>
      <span>
        用户名
        <i nz-icon nz-tooltip nzTitle="昵称长度不超过 6 位" type="question-circle"
theme="outline"></i>
      </span>
    </nz-form-label>
    <nz-form-control [nzSm]="14" [nzXs]="24">
      <input nz-input id="nickname" formControlName="nickname">
      <nz-form-explain *ngIf="validateForm.get('nickname').dirty &&
validateForm.get('nickname').errors">请输入用户名!</nz-form-explain>
    </nz-form-control>
  </nz-form-item>
  <nz-form-item>
    <nz-form-label [nzSm]="6" [nzXs]="24" nzRequired>生日</nz-form-label>
    <nz-form-control [nzSm]="14" [nzXs]="24">
      <nz-date-picker formControlName="birthday"></nz-date-picker>
      <nz-form-explain *ngIf="validateForm.get('birthday').dirty &&
validateForm.get('birthday').errors">请选择生日</nz-form-explain>
    </nz-form-control>
  </nz-form-item>
  <nz-form-item>
    <nz-form-label [nzSm]="6" [nzXs]="24" nzRequired nzFor="email">电子邮箱
</nz-form-label>
    <nz-form-control [nzSm]="14" [nzXs]="24">
      <input nz-input formControlName="email" id="email">
      <nz-form-explain *ngIf="validateForm.get('email').dirty &&
validateForm.get('email').errors">请输入电子邮箱!</nz-form-explain>
    </nz-form-control>
  </nz-form-item>
  <nz-form-item>
    <nz-form-label [nzSm]="6" [nzXs]="24" nzFor="phoneNumber" nzRequired>手
机号</nz-form-label>
    <nz-form-control [nzSm]="14" [nzXs]="24"
[nzValidateStatus]="validateForm.controls['phoneNumber']">
      <nz-input-group [nzAddOnBefore]="addOnBeforeTemplate">
        <ng-template #addOnBeforeTemplate>
          <nz-select formControlName="phoneNumberPrefix" style="width: 70px;">
            <nz-option nzLabel="+86" nzValue="+86"></nz-option>
```

```
            <nz-option nzLabel="+87" nzValue="+87"></nz-option>
          </nz-select>
        </ng-template>
        <input formControlName="phoneNumber" id="'phoneNumber'" nz-input>
      </nz-input-group>
      <nz-form-explain *ngIf="validateForm.get('phoneNumber').dirty &&
validateForm.get('phoneNumber').errors">请输入手机号!</nz-form-explain>
    </nz-form-control>
  </nz-form-item>
  <nz-form-item nz-row style="margin-bottom:8px;">
    <nz-form-control [nzSpan]="14" [nzOffset]="6">
      <button nz-button nzType="primary">保存</button>
    </nz-form-control>
  </nz-form-item>
</form>
```

【代码解析】由于是表单的展示，所以内容比较多。首先是一个用户名，直接输入文本，限制不超过 6 位。之后是生日、电子邮箱与手机号，都是必填项。如果没有通过条件，nz-form-explain 就会随时被“响应”出来提示错误信息。

接下来编写最关键的部分的代码，表单验证基本就在 TS 文件中了，form-test.component. ts 的代码如下：

```
import { Component, OnInit } from '@angular/core';
import { FormBuilder, FormGroup, Validators } from '@angular/forms';

@Component({
  selector: 'app-form-test',
  templateUrl: './form-test.component.html',
  styleUrls: ['./form-test.component.css']
})
export class FormTestComponent implements OnInit {
  validateForm: FormGroup;

  constructor(private fb: FormBuilder) {
  }

  ngOnInit(): void {
    this.validateForm = this.fb.group({
      nickname: [null, [Validators.required, Validators.maxLength(6)]],
      birthday: [null, [Validators.required]],
      email: [null, [Validators.email, Validators.required]],
      phoneNumberPrefix: ['+86'],
      phoneNumber: [null, [Validators.required,
Validators.pattern(/^1([38]\d|5[0-35-9]|7[3678])\d{8}$/)]]
    });
  }

  submitForm(): void {
    let params = {};
    for (const i in this.validateForm.controls) {
      this.validateForm.controls[i].markAsDirty();
      this.validateForm.controls[i].updateValueAndValidity();
      if (!(this.validateForm.controls[i].status == 'VALID') &&
this.validateForm.controls[i].status !== 'DISABLED') {
```

```
        return;
      }
      if (this.validateForm.controls[i] &&
this.validateForm.controls[i].value) {
        params[i] = this.validateForm.controls[i].value;
      } else {
        params[i] = '';
      }
    }
    this.setDate('birthday');
    params['birthday'] = this.validateForm.get('birthday').value;
    console.log(params);
  }

    // 设置日期格式
    setDate(dates) {
      const time = new Date(this.validateForm.get(dates).value);
      const datetime = time.getFullYear() + '-' +
this.formatDayAndMonth(time.getMonth() + 1) + '-' +
this.formatDayAndMonth(time.getDate());
      this.validateForm.get(dates).setValue(datetime);
    }

    formatDayAndMonth(val) {
      if (val < 10) {
        val = '0' + val;
      }
      return val;
    }
}
```

【代码解析】我们通过新建一个 FormGroup 类型的参数，在进入页面时初始化，设置它的各种必填项和验证规则。在提交的时候，再循环检查是否全部通过，没有通过的话，直接显示错误提示。全部通过后，我们将所带的参数放到一个对象里提交即可。程序的运行结果如图 7.6 所示。

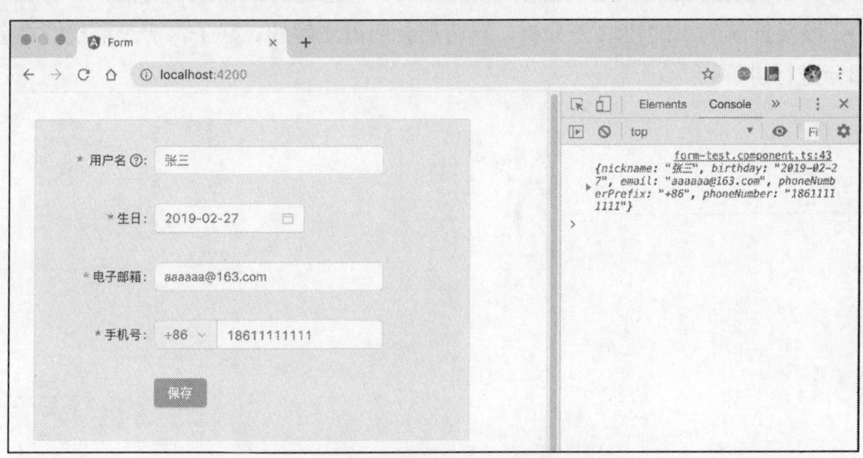

图 7.6 所有验证通过后的屏幕显示结果

这个时候我们将所填写的内容都删掉，由于是响应式表单，所以不需要单击“提交”按钮，错误信息会直接弹出，如图 7.7 所示。

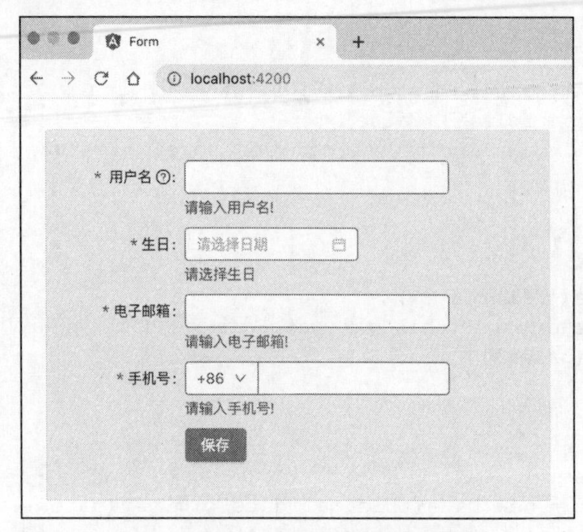

图 7.7　验证未通过，弹出提示信息

如果显示结果没问题的话，这个响应式表单的内容就算是完成了。至此，读者们已经尝试过这两种表单的验证过程，对它们的使用方法、优势应该也有了自己的理解，所以在之后的开发过程中，可以根据需要灵活地使用。

7.4 小 结

本章主要讲解了两种不同的表单验证方式：响应式表单大而全面，可扩展性强；模板驱动型表单则小巧灵活，更适用于简单的页面。对如何使用和选择，本章依然是通过两个不同的实战项目，带领读者了解了它们不同的用法。在开发完这两个实战例子之后，相信读者对表单有了更深入的理解。因为不管多么复杂的表单，也只是输入项较多，表单验证的原理是不变的，所以面对复杂表单时不必担心，慢慢拆解为小的模块去分析，一切都会迎刃而解的。好了，下一章我们将会对十分重要的路由部分进行讲解。

第8章 路由

在 Angular 中，它的 Router（路由器）并不在@angular/core 中，它拥有自己的包@angular/router，在导入的时候需要注意。

本章主要涉及的知识点有：

- 路由的基本用法
- 路由的位置策略
- 路由的跳转与传参
- 路由守卫
- 嵌套路由
- 实战练习：路由框架的搭建

8.1 路由的基本用法

路由实际上是 URL 与页面或者说内容的映射。比如在路由中设置了/login 为登录页面后，在域名后输入/login 就能跳转到登录页面。这个时候我们在按钮的单击事件上增加了路由的跳转，就可以实现页面的跳转了。

8.1.1 路由的配置

【示例 8-1】接下来介绍路由的基本用法，输入以下指令新建一个项目，并导入 ng-zorro。

```
ng new router
cd router
ng add ng-zorro-antd
```

首先新建一个 app-routing.module.ts，并输入以下代码：

```
import { NgModule } from '@angular/core';
import { Routes, RouterModule } from '@angular/router';

const routes: Routes = [];

@NgModule({
  imports: [RouterModule.forRoot(routes)],
  exports: [RouterModule]
})
export class AppRoutingModule { }
```

【代码解析】routes 目前是一个空的数组，它会在 imports 中被 RouterModule 注册。等我们有了组件后，将组件写到里面就可以执行了。

接下来在 app.module.ts 引入我们的路由配置。

```
import { BrowserModule } from '@angular/platform-browser';
import { NgModule } from '@angular/core';

import { AppRoutingModule } from './app-routing.module';
import { AppComponent } from './app.component';

@NgModule({
  declarations: [
    AppComponent
  ],
  imports: [
    BrowserModule,
    AppRoutingModule
  ],
  providers: [],
  bootstrap: [AppComponent]
})
export class AppModule { }
```

最后在 app.component.html 输入以下代码：

```
<router-outlet></router-outlet>
```

8.1.2 让路由与组件对应

【示例 8-2】接下来新建三个组件，作为路由功能的演示。

```
ng g component a --spec=false
ng g component b --spec=false
ng g component c --spec=false
```

将 app-routing.module.ts 修改为以下代码：

```
import { NgModule } from '@angular/core';
import { Routes, RouterModule } from '@angular/router';
import { AComponent } from './a/a.component';
import { BComponent } from './b/b.component';
```

```
import { CComponent } from './c/c.component';

const routes: Routes = [
  { path: 'A', component: AComponent },
  { path: 'B', component: BComponent },
  { path: 'C', component: CComponent }
];

@NgModule({
  imports: [RouterModule.forRoot(routes)],
  exports: [RouterModule]
})
export class AppRoutingModule { }
```

【代码解析】path 与 component 就是路径与组件的对应关系，设置完之后，在路径后加上/path 即可看到对应的组件，如图 8.1 所示。

图 8.1　组件与 URL 对应

最后修改 app.component.html 与 app.component.css 来设置几个按钮，实现按钮控制不同路由展示的功能。

```
// app.component.html
<div style="margin:15px;">
  <button nz-button nzType="primary" [routerLink]="['/A']">A 组件</button>
  <button nz-button nzType="primary" [routerLink]="['/B']">B 组件</button>
  <button nz-button nzType="primary" [routerLink]="['/C']">C 组件</button>
  <router-outlet></router-outlet>
</div>

// app.component.css
[nz-button] {
  margin-right: 8px;
  margin-bottom: 12px;
}
```

【代码解析】我们为三个按钮分别设置了不同的 routerLink，用来跳转到不同的页面，如图 8.2 所示。关于跳转，会在后面的小节中进行详细讲解。

图 8.2　使用按钮控制路由跳转

8.1.3　设置默认路径

以我们的程序为例，刚打开一个页面，默认的路径为 http://localhost:4200/，所以在三个按钮下面是一片空白，如图 8.3 所示。

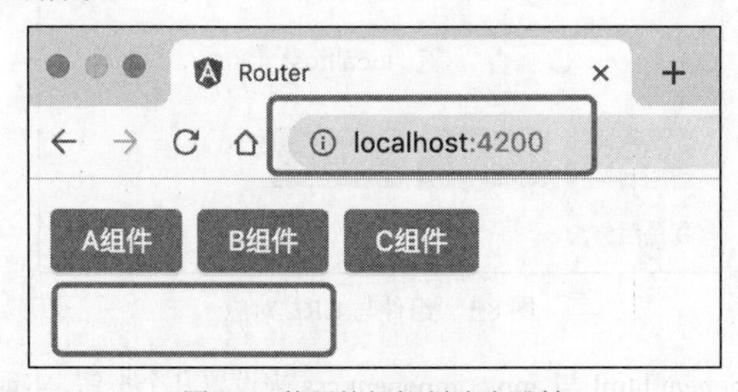

图 8.3　使用按钮控制路由跳转

【示例 8-3】如果想做到在默认路径时，显示某个组件该怎么做呢？其实很简单，只要对 app-routing.module.ts 进行简单的修改即可。

```
import { NgModule } from '@angular/core';
import { Routes, RouterModule } from '@angular/router';
import { AComponent } from './a/a.component';
import { BComponent } from './b/b.component';
import { CComponent } from './c/c.component';

const routes: Routes = [
  { path: '', component: AComponent},
  { path: 'A', component: AComponent },
  { path: 'B', component: BComponent },
  { path: 'C', component: CComponent }
];

@NgModule({
  imports: [RouterModule.forRoot(routes)],
  exports: [RouterModule]
})
export class AppRoutingModule { }
```

【代码解析】在 routes 添加了空的 path，并设置为 AComponent，这样在 URL 为

http://localhost:4200/的时候也能默认显示 A 组件了。结果如图 8.4 所示。

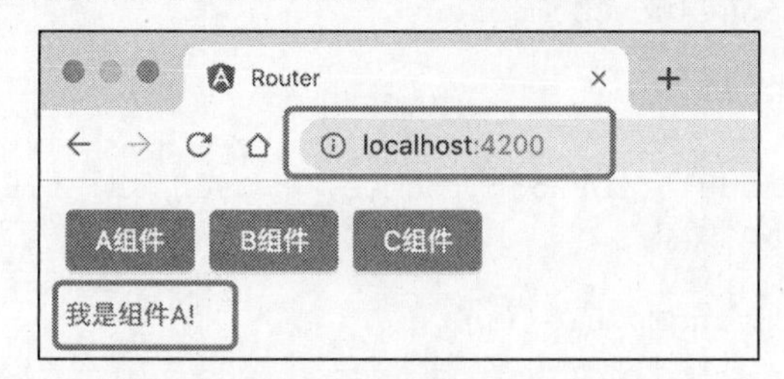

图 8.4　使用按钮控制路由跳转——增加了默认路径

8.2　路由的位置策略

Angular 路由的位置策略（LocationStrategy）有两种：

- PathLocationStrategy：默认的策略，支持 HTML 5 pushState 风格。
- HashLocationStrategy：支持 hash URL 风格。

在 Angular 中，只有小部分的项目需要用到 HashLocationStrategy，大部分时候则使用默认的 PathLocationStrategy 风格。因而我们在下面的小节主要分析 HashLocationStrategy，并在最后做一下比较。

8.2.1　HashLocationStrategy

当路由器导航到一个新的组件时，它的视图的 URL 其实是本地的，浏览器不会把该 URL 发给服务器，并且不会重新加载此页面。旧的浏览器在当前地址的 URL 变化时总会往服务器发送页面请求。其中的例外规则是：当这些地址变化位于“#”符号（被称为“hash”）后面时不会发送。通过把应用内的路由 URL 拼接在#之后，路由器就可以获得这条“例外规则”带来的优点。

通过修改路由的位置策略，就可以用 URL 片段“#”代替 history API，以起到防止浏览器向服务器发送页面请求。下一节会讲到它的使用方法。

8.2.2　如何使用位置策略

【示例 8-4】使用的方式很简单，只需要在 app.routing.module.ts 中给 forRoot 多传一个参数。

```
import { NgModule } from '@angular/core';
import { Routes, RouterModule } from '@angular/router';
import { AComponent } from './a/a.component';
import { BComponent } from './b/b.component';
import { CComponent } from './c/c.component';

const routes: Routes = [
  { path: '', component: AComponent },
  { path: 'A', component: AComponent },
```

```
  { path: 'B', component: BComponent },
  { path: 'C', component: CComponent }
];

@NgModule({
  imports: [RouterModule.forRoot(routes, { useHash: true })],
  exports: [RouterModule]
})
export class AppRoutingModule { }

import { NgModule } from '@angular/core';
import { Routes, RouterModule } from '@angular/router';
import { AComponent } from './a/a.component';
import { BComponent } from './b/b.component';
import { CComponent } from './c/c.component';

const routes: Routes = [
  { path: '', component: AComponent },
  { path: 'A', component: AComponent },
  { path: 'B', component: BComponent },
  { path: 'C', component: CComponent }
];

@NgModule({
  imports: [RouterModule.forRoot(routes, { useHash: true })],
  exports: [RouterModule]
})
export class AppRoutingModule { }
```

【代码解析】在 forRoot 中直接添加 useHash 并设置为 true 即可使用 HashLocationStrategy，否则默认为不带“#”的 PathLocationStrategy。该程序的执行结果如图 8.5 所示。

图 8.5 路由的位置策略

8.2.3 如何选择两种位置策略

在上一节已经提到过了 HashLocationStrategy 的特性，我们知道它是在为了兼容旧版浏览器而设的。选择方式其实很简单，我们以使用默认 HTML 5 风格的 PathLocationStrategy 为主，只在不得已而需要兼容旧版时再使用 HashLocationStrategy。

8.3 路由的跳转与传参

使用路由来完成了URL与组件的映射后，下一步就需要考虑如何实现跳转和传参。下面就来分别介绍不同方式的路由跳转与传参。

8.3.1 路由的跳转

首先回顾一下8.1节中app.component.html文件展示的一种跳转方式。

```
// app.component.html
<div style="margin:15px;">
  <button nz-button nzType="primary" [routerLink]="['/A']">A 组件</button>
  <button nz-button nzType="primary" [routerLink]="['/B']">B 组件</button>
  <button nz-button nzType="primary" [routerLink]="['/C']">C 组件</button>
  <router-outlet></router-outlet>
</div>
```

【代码解析】routerLink这个属性所提供的功能就是在应用中链接到指定的路由。方括号的写法属于绑定，然后用单引号进行了转义。如果路由的位置是固定的，以组件B为例，那也可以用以下方式来编写。

```
// app.component.html
<div style="margin:15px;">
  <button nz-button nzType="primary" [routerLink]="['/A']">A 组件</button>
  <button nz-button nzType="primary" routerLink="/B">B 组件</button>
  <button nz-button nzType="primary" [routerLink]="['/C']">C 组件</button>
  <router-outlet></router-outlet>
</div>
```

【示例8-5】接下来修改app.component.html以实现另一种方式的路由跳转。

```
// app.component.html
<div style="margin:15px;">
  <button nz-button nzType="primary" [routerLink]="['/A']">A 组件</button>
  <button nz-button nzType="primary" routerLink="/B">B 组件</button>
  <button nz-button nzType="primary" (click)="routerToC()">C 组件</button>
  <router-outlet></router-outlet>
</div>
```

【代码解析】修改了组件C，去掉了routerLink，并添加了一个单击事件，之后将在app.component.ts中实现路由的跳转。

```
// app.component.ts
import { Component } from '@angular/core';
import { Router } from '@angular/router';

@Component({
  selector: 'app-root',
  templateUrl: './app.component.html',
  styleUrls: ['./app.component.css']
```

```
})
export class AppComponent {
  title = 'router';

  constructor(private router: Router) {
  }

  routerToC() {
    this.router.navigate(['/C']);
  }
}
```

【代码解析】首先从@angular/router 导入路由器并在构造函数中进行依赖注入。之后为打算跳转到组件 C 的按钮添加单击事件。方式也很简单，使用 router 中的 navigate 方法即可。因为显示结果与之前的一样，这里就不提供示例图了。

总结一下两种路由跳转方式的使用方法。

在 HTML 文件中使用指令进行跳转。

```
routerLink="跳转路径"
```

在 TS 文件中引入 Router（路由器）进行跳转。

```
this.router.navigate(['跳转路径']);
```

8.3.2 路由的传参

为了便于演示，再新建三个组件，作为传值的演示。分别为学生、教师和家长组件。

```
ng g component student --spec=false
ng g component teacher --spec=false
ng g component parent --spec=false
```

在 app-routing.module.ts 中将它们导入并设置对应的 path。

```
import { NgModule } from '@angular/core';
import { Routes, RouterModule } from '@angular/router';
import { AComponent } from './a/a.component';
import { BComponent } from './b/b.component';
import { CComponent } from './c/c.component';
import { StudentComponent } from './student/student.component';
import { TeacherComponent } from './teacher/teacher.component';
import { ParentComponent } from './parent/parent.component';

const routes: Routes = [
  { path: '', component: AComponent },
  { path: 'A', component: AComponent },
  { path: 'B', component: BComponent },
  { path: 'C', component: CComponent },
  { path: 'student', component: StudentComponent },
  { path: 'teacher', component: TeacherComponent },
  { path: 'parent', component: ParentComponent }
];
```

```
@NgModule({
  imports: [RouterModule.forRoot(routes, { useHash: true })],
  exports: [RouterModule]
})
export class AppRoutingModule { }
```

首先需要把三个组件对应的选项给构建出来，还是使用 NG-ZORRO 的组件来实现，修改 app.component.html 和 app.component.ts 中的代码如下：

```
// app.component.html
<div style="margin:15px;">
  <button nz-button nzType="primary" [routerLink]="['/A']">A 组件</button>
  <button nz-button nzType="primary" routerLink="/B">B 组件</button>
  <button nz-button nzType="primary" (click)="routerToC()">C 组件</button>
  <div>
    <nz-select style="width: 120px;" [(ngModel)]="selectedValue"
(ngModelChange)="valueChange()" nzPlaceHolder="请选择">
      <nz-option nzValue="student" nzLabel="学生组件"></nz-option>
      <nz-option nzValue="teacher" nzLabel="教师组件"></nz-option>
      <nz-option nzValue="parent" nzLabel="家长组件"></nz-option>
    </nz-select>
  </div>
  <router-outlet></router-outlet>
</div>

// app.component.ts
import { Component } from '@angular/core';
import { Router } from '@angular/router';

@Component({
  selector: 'app-root',
  templateUrl: './app.component.html',
  styleUrls: ['./app.component.css']
})
export class AppComponent {
  title = 'router';
  selectedValue = '';

  constructor(private router: Router) {
  }

  routerToC() {
    this.router.navigate(['/C']);
  }

  valueChange() {
    if (this.selectedValue === 'student') {
      this.router.navigate(['/student']);
    } else if (this.selectedValue === 'teacher') {
      this.router.navigate(['/teacher']);
    } else {
      this.router.navigate(['/parent']);
    }
  }
```

```
}
```

【代码解析】这里使用 Select 组件实现了路由的选择，在 valueChange 方法中根据选中的值来判断显示的路由组件。

【示例 8-6】接下来讲解第一种传参方式 queryParams，这里直接把三个文件需要修改的代码列出，一起进行讲解。

```
// app.component.ts
...
export class AppComponent {
  ...
  valueChange() {
    if (this.selectedValue === 'student') {
      this.router.navigate(['/student'], { queryParams: { name: '张三' } });
    } else if (this.selectedValue === 'teacher') {
      this.router.navigate(['/teacher']);
    } else {
      this.router.navigate(['/parent']);
    }
  }
}

// student.component.html
<p>
  学生名：{{name}}!
</p>

// student.component.ts
import { Component, OnInit } from '@angular/core';
import { ActivatedRoute } from '@angular/router';

@Component({
  selector: 'app-student',
  templateUrl: './student.component.html',
  styleUrls: ['./student.component.css']
})
export class StudentComponent implements OnInit {
  name: '';

  constructor(private activatedRoute: ActivatedRoute) { }

  ngOnInit() {
    this.name = this.activatedRoute.snapshot.queryParams['name'];
  }
}
```

【代码解析】首先修改路由跳转的方式，在参数后面增加 queryParams，这时跳转后就可以看到链接的尾部多出了“?name=张三”的字样，说明传值已经成功。最后在 student 组件中，使用 ActivatedRoute 获取上一个页面传来的值，并赋值给 name 参数，传值就大功告成了。程序执行结果如图 8.6 所示。如果要在 HTML 文件中进行传值，可以在标签中加入以下代码进行传值，结果都是一样的。后面的例子就不再列举。

```
[queryParams]="{name:'张三'}"
```

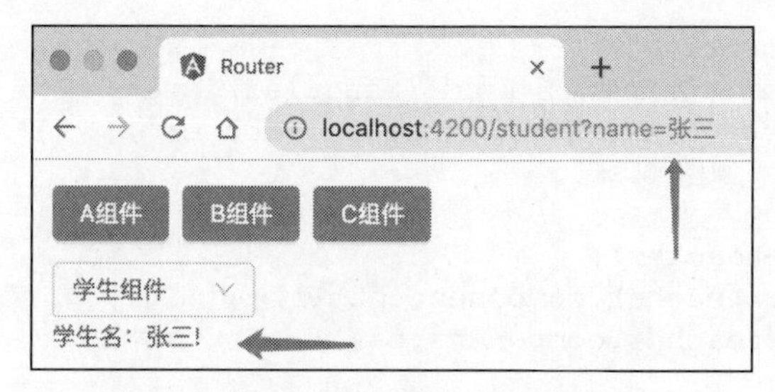

图 8.6　使用 queryParams 进行路由传值

【示例 8-7】接下来讲解第二种传参方式，它与上一种方式不同，需要在路由文件中对 path 先进行配置才能生效，否则直接传值会报错。先修改 app-routing.module.ts 的代码。

```
// app-routing.module.ts
...
const routes: Routes = [
  { path: '', component: AComponent },
  { path: 'A', component: AComponent },
  { path: 'B', component: BComponent },
  { path: 'C', component: CComponent },
  { path: 'student', component: StudentComponent },
  { path: 'teacher/:name', component: TeacherComponent },
  { path: 'parent', component: ParentComponent }
];

@NgModule({
  imports: [RouterModule.forRoot(routes)],
  exports: [RouterModule]
})
export class AppRoutingModule { }
```

接下来修改这三个文件的代码以实现使用 path 进行路由传值与展示，大体上与上一种方式很相似，但是在细微之处还是有所区别。

```
// app.component.ts
...
export class AppComponent {
  ...
  valueChange() {
    if (this.selectedValue === 'student') {
      this.router.navigate(['/student'], { queryParams: { name: '张三' } });
    } else if (this.selectedValue === 'teacher') {
      this.router.navigate(['/teacher', '李四']);
    } else {
      this.router.navigate(['/parent']);
    }
  }
}

// teacher.component.html
<p>
  教师名：{{name}}!
</p>
```

```
// teacher.component.ts
import { Component, OnInit } from '@angular/core';
import { ActivatedRoute } from '@angular/router';

@Component({
  selector: 'app-teacher',
  templateUrl: './teacher.component.html',
  styleUrls: ['./teacher.component.css']
})
export class TeacherComponent implements OnInit {
  name: '';

  constructor(private activatedRoute: ActivatedRoute) { }

  ngOnInit() {
    this.name = this.activatedRoute.snapshot.params['name'];
  }
}
```

【代码解析】首先注意一下传值，是直接在路由地址后传送了内容，而不是使用 queryParams 再进行传值。这种方式传值后，在 URL 中不会有问号与等号，而是直接将内容拼接在里面。接收的方式也改为了 params['name']。程序的执行结果如图 8.7 所示。

图 8.7　使用 path 进行路由传值

【示例 8-8】最后一种方式比较简单，应用场景稍微少一些，直接在路由的配置文件中，增加 data 参数即可进行传值。

```
// app-routing.module.ts
...
const routes: Routes = [
  { path: '', component: AComponent },
  { path: 'A', component: AComponent },
  { path: 'B', component: BComponent },
  { path: 'C', component: CComponent },
  { path: 'student', component: StudentComponent },
  { path: 'teacher/:name', component: TeacherComponent },
  { path: 'parent', component: ParentComponent, data: { name: '王五' } }
];

@NgModule({
  imports: [RouterModule.forRoot(routes)],
  exports: [RouterModule]
})
export class AppRoutingModule { }
```

接下来从家长组件取值并显示出来即可。

```
// parent.component.html
<p>
  家长名：{{name}}!
</p>

// parent.component.ts
import { Component, OnInit } from '@angular/core';
import { ActivatedRoute } from '@angular/router';

@Component({
  selector: 'app-parent',
  templateUrl: './parent.component.html',
  styleUrls: ['./parent.component.css']
})
export class ParentComponent implements OnInit {
  name: '';

  constructor(private activatedRoute: ActivatedRoute) { }

  ngOnInit() {
    this.name = this.activatedRoute.snapshot.data['name'];
  }
}
```

【代码解析】取值方式改为从 snapshot 的 data 属性中取值，与前两种最大的区别是：在 URL 中不会显示出所传的参数。程序执行的结果如图 8.8 所示。

图 8.8 使用 data 进行路由传值

8.4 子路由

在使用路由的过程中，嵌套多层子路由是十分常见的，在 Angular 中，使用方法十分简单。我们将上一小节中的学生、教师、家长组件作为子路由，放置到 C 组件中作为例子来说明一下。

【示例 8-9】首先将这三个组件从根组件迁移到组件 C 中去。根组件中需要删除的内容不再列出，以下只展示组件 C 中的代码。

```
// c.component.html
<p>
  我是组件 C!
```

```
</p>

<div>
  <nz-select style="width: 120px;" [(ngModel)]="selectedValue"
(ngModelChange)="valueChange()" nzPlaceHolder="请选择">
    <nz-option nzValue="student" nzLabel="学生组件"></nz-option>
    <nz-option nzValue="teacher" nzLabel="教师组件"></nz-option>
    <nz-option nzValue="parent" nzLabel="家长组件"></nz-option>
  </nz-select>
</div>

// c.component.ts
import { Component, OnInit } from '@angular/core';
import { Router } from '@angular/router';

@Component({
  selector: 'app-c',
  templateUrl: './c.component.html',
  styleUrls: ['./c.component.css']
})
export class CComponent implements OnInit {
  selectedValue = '';
  constructor(private router: Router) { }

  ngOnInit() {
  }
  valueChange() {
    if (this.selectedValue === 'student') {
      this.router.navigate(['/C/student'], { queryParams: { name: '张三' } });
    } else if (this.selectedValue === 'teacher') {
      this.router.navigate(['/C/teacher', '李四']);
    } else {
      this.router.navigate(['/C/parent']);
    }
  }
}
```

【代码解析】大多数代码都是移植过来的，但是需要注意一点，navigate 中的路径前多了一个 /C，这就说明它要路由到 C 的子路由中去。

最后将三个组件配置为子路由即可完成，代码如下：

```
// app-routing.module.ts
...
const routes: Routes = [
  { path: '', component: AComponent },
  { path: 'A', component: AComponent },
  { path: 'B', component: BComponent },
  {
    path: 'C',
    component: CComponent,
    children: [
      { path: 'student', component: StudentComponent },
      { path: 'teacher/:name', component: TeacherComponent },
      { path: 'parent', component: ParentComponent, data: { name: '王五' } }
```

```
    ]
  }
];

@NgModule({
  imports: [RouterModule.forRoot(routes)],
  exports: [RouterModule]
})
export class AppRoutingModule { }
```

【代码解析】在配置的路由中增加 children 即可设置子路由，它与普通的路由一样，也是传递一个数组作为参数。如果需要叠加多层的路由，也是增加 children 即可实现。程序的执行结果如图 8.9 所示。

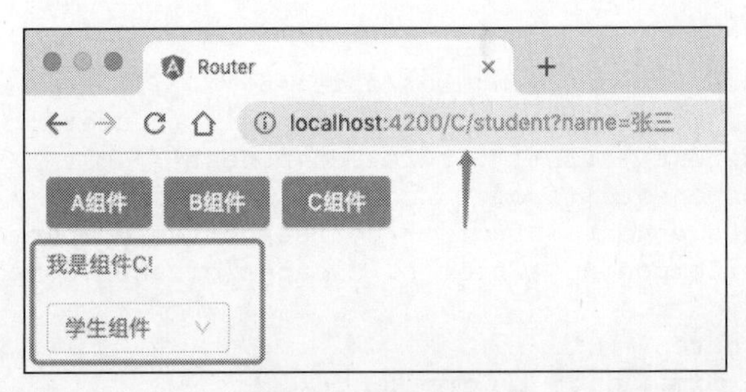

图 8.9　设置子路由

8.5　实战练习：路由框架的搭建

在制作网站时，我们几乎都需要使用到路由技术。本节项目实战将展示一个比较常见的路由框架的实现，如图 8.10 所示。

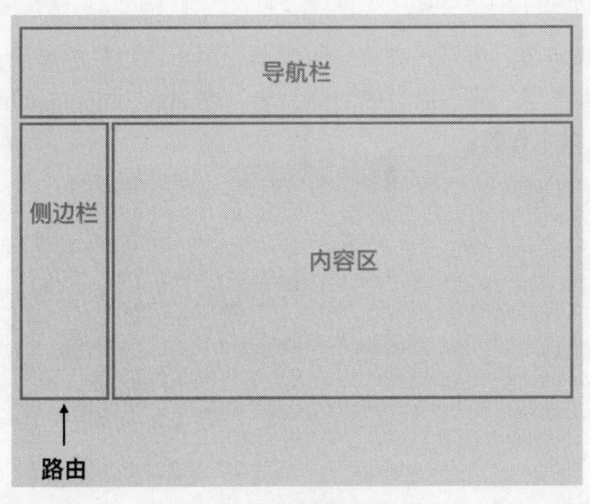

图 8.10　常见路由框架

首先新建一个项目，并添加 NG-ZORRO。

```
ng new layout
```

```
cd layout
ng add ng-zorro-antd
```

然后根据我们的设计，给导航栏与侧边栏添加对应的内容。导航栏用来放商品类型，侧边栏放商品名，内容区放具体商品的图片与文字。首先新建几种商品，输入以下代码来新建这些商品组件。

```
ng g component tv--spec=false
ng g component cola --spec=false
ng g component coffee --spec=false
ng g component cleaner --spec=false
```

添加完了组件之后，首先要做的就是为它们设置对应的路由，在 app-routing.module.ts 输入以下代码：

```
import { NgModule } from '@angular/core';
import { Routes, RouterModule } from '@angular/router';
import { TvComponent } from './tv/tv.component';
import { ColaComponent } from './cola/cola.component';
import { CoffeeComponent } from './coffee/coffee.component';
import { CleanerComponent } from './cleaner/cleaner.component';

const routes: Routes = [
  { path: 'tv', component: TvComponent},
  { path: 'cola', component: ColaComponent},
  { path: 'coffee', component: CoffeeComponent},
  { path: 'cleaner', component: CleanerComponent}
];

@NgModule({
  imports: [RouterModule.forRoot(routes)],
  exports: [RouterModule]
})
export class AppRoutingModule { }
```

这四个组件分别代表电视、可乐、咖啡与吸尘器，用于侧边栏的展示。导航栏中则放入热销、酒水饮料、生活物品三大类型，并添加对应的商品名。在 app.component.html、app.component.scss 和 app.component.ts 输入以下代码：

```
// app.component.html
<nz-layout class="layout">
  <nz-header>
    <ul nz-menu [nzTheme]="'dark'" [nzMode]="'horizontal'" style="line-height:
64px;">
      <li nz-menu-item [nzSelected]="true" (click)="hot()">热销</li>
      <li nz-menu-item (click)="drinks()">酒水饮料</li>
      <li nz-menu-item (click)="livingGoods()">生活用品</li>
    </ul>
  </nz-header>
  <nz-layout>
    <nz-sider [nzWidth]="200" style="background:#fff">
      <ul nz-menu [nzMode]="'inline'" style="height:100%">
        <li *ngFor="let item of items" nz-menu-item
[routerLink]="[item.path]">{{item.name}}</li>
```

```
      </ul>
    </nz-sider>
    <nz-content class="content">
      <router-outlet></router-outlet>
    </nz-content>
  </nz-layout>
</nz-layout>

// app.component.scss
.layout {
    height: 100%;
}

.content {
    padding: 30px;
}

// app.component.ts
import { Component } from '@angular/core';

@Component({
  selector: 'app-root',
  templateUrl: './app.component.html',
  styleUrls: ['./app.component.scss']
})
export class AppComponent {
  title = 'layout';
  items: Array<Object> = [];
  constructor() {
    this.hot();
  }

  hot() {
    this.items = [
      { name: '电视', path: 'tv' },
      { name: '可乐', path: 'cola' },
      { name: '咖啡', path: 'coffee' }
    ];
  }

  drinks() {
    this.items = [
      { name: '可乐', path: 'cola' },
      { name: '咖啡', path: 'coffee' }
    ];
  }

  livingGoods() {
    this.items = [
      { name: '电视', path: 'tv' },
      { name: '吸尘器', path: 'cleaner' }
    ];
  }
}
```

【代码解析】我们分别把热销、酒水饮料、生活用品设置了不同的侧边栏列表内容，并根据它们的单击事件进行切换。侧边栏则通过 ngFor 显示出列表，并绑定路由对应的 path。其中 items 这个数组，则是我们存放侧边栏数据的变量，name 为商品名，path 为对应路径。

最后以电视为例，添加一个商品进去以便展示。其他类型的就不再重复演示，原理都是一样的，我们主要的目的是展示路由功能。在 tv.component.html 输入以下代码：

```
<img style="width:200px;height:150px" src="./assets/tv.jpg">
<br>
<span>电视大降价，赶快购买吧！</span>
```

最后，我们就可以看到如图 8.11 所示的程序执行结果，对于程序中的商品图片，读者也可以根据自己的需要进行替换。

图 8.11　常见路由框架

8.6　小　结

本章主要讲解了路由的基本用法、位置策略、跳转和传参等知识点，并在最后做了一个实战练习。路由是十分常见并且重要的知识点，本章内容较多，因而需要多加练习方能掌握好。下一章我们将对 Angular 的测试部分进行讲解。

第9章

Angular 中的测试

在软件开发的流程中，测试往往容易被轻视。在一些中小公司，甚至没有一个完善的测试流程。想要产品保持稳定的质量，测试的重要性怎么强调也不为过。在 Angular 中，官方为我们提供了很多优秀的测试工具，方便大家轻松地完成测试相关代码。本章将着重介绍测试工具的相关知识与用法。

本章主要涉及的知识点有：

- 测试的意义
- 单元测试
- Angular 测试工具
- 端对端测试

9.1 测试的意义

有读者可能有疑问，为什么有了测试人员还需要编写单元测试？我们来分析一下自己编写测试单元的几个意义。

（1）提前发现问题，提前暴露问题给开发人员，免去了与测试人员的沟通、交流成本。

（2）提高代码质量、可维护性，为之后的版本迭代打下基础。

（3）自动化执行，免去一些重新性的测试操作，一劳永逸。

以上三点也可以说是测试的优点。当然，测试代码也不一定是必须要写的。根据我们项目的具体场景与进度，如果觉得有必要的话，就继续学习一下如何使用 Angular 来编写测试吧。

9.2 第一个测试例子

单元测试也称为模块测试，用于对程序单元进行正确性测试。程序单元是最小的可测试模块，单元可以理解为一个函数、一个类或者一个窗口。

【示例 9-1】新建一个项目，来运行第一个单元测试。

```
ng new ngtest
ng add ng-zorro-antd
```

等待 NPM 安装完毕后，直接使用以下指令进行测试。不出意外的话，会遇到如图 9.1 所示的错误。

```
ng test
```

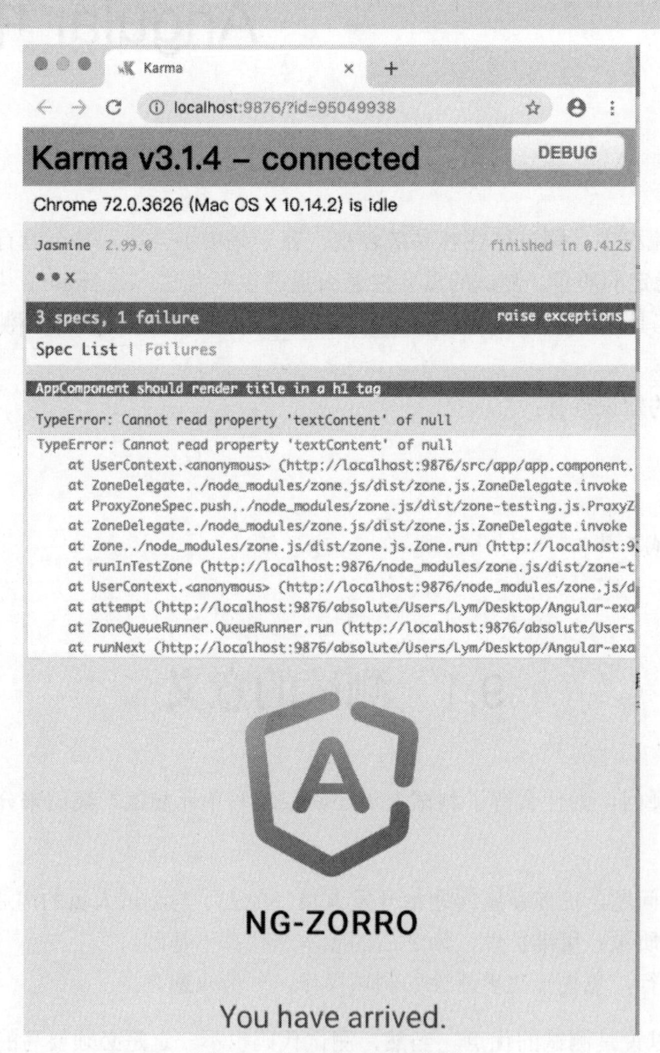

图 9.1 执行单元测试报错

为什么我们对刚生成的默认项目执行单元测试，也会报错呢。分析以下原因，在新项目生成后，会自动生成一个 app.component.spec.ts 文件，里面已经编写好了单元测试，代码如下：

```
import { TestBed, async } from '@angular/core/testing';
import { AppComponent } from './app.component';

describe('AppComponent', () => {
  beforeEach(async(() => {
    TestBed.configureTestingModule({
      declarations: [
        AppComponent
      ],
    }).compileComponents();
  }));

  it('should create the app', () => {
    const fixture = TestBed.createComponent(AppComponent);
    const app = fixture.debugElement.componentInstance;
    expect(app).toBeTruthy();
  });

  it(`should have as title 'ngtest'`, () => {
    const fixture = TestBed.createComponent(AppComponent);
    const app = fixture.debugElement.componentInstance;
    expect(app.title).toEqual('ngtest');
  });

  it('should render title in a h1 tag', () => {
    const fixture = TestBed.createComponent(AppComponent);
    fixture.detectChanges();
    const compiled = fixture.debugElement.nativeElement;
    expect(compiled.querySelector('h1').textContent).toContain('Welcome to
ngtest!');
  });
});
```

【代码解析】这里先分析一下这些代码的作用是什么。其中第一部分是把这个测试用例绑定到 AppComponent 上了，说明它是用来测试 AppComponent 的。之后有三个单元测试，第一个用于测试是否创建了 AppComponent；第二个用于测试浏览器的标题是否为 ngtest，也就是我们项目的名字；第三个用于测试是否存在 h1 标签，并且其是否内容为“Welcome to ngtest!”。之后再对比一下图 9.1 中的报错信息，我们就可以知道是第三个测试用例出错了。

为什么会出现这个错误呢？看一下 app.component.html，原因是安装了 NG-ZORRO 之后这个文件被覆盖了，所以那个 h1 标签没了。修改 app.component.html 添加上 h1 标签，看看是否能解决报错问题。

```
<h1>Welcome to ngtest!</h1>
<a href="https://github.com/NG-ZORRO/ng-zorro-antd" target="_blank"
style="display: flex;align-items: center;justify-content: center;height:
100%;width: 100%;">
  <img height="300"
src="https://img.alicdn.com/tfs/TB1NvvIwTtYBeNjy1XdXXXXyVXa-89-131.svg">
</a>
```

重新执行测试指令，可以看到所有的单元测试都可以通过了，如图 9.2 所示。

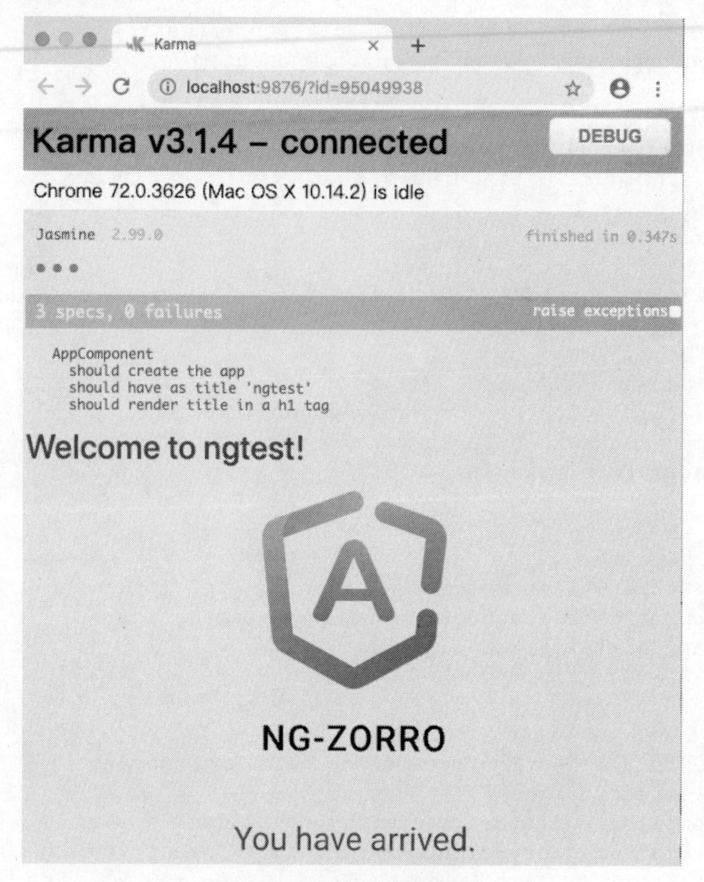

图 9.2 单元测试全部通过

9.3 Angular 测试工具

从前面的例子其实就可以看出，实际上在使用 Angular CLI 构建项目的过程中，它会自动帮我们生成单元测试的相关工具并配置的相关文件。比如 app.component.spec.ts、karma.conf.js、test.ts 等。

9.3.1 Jasmine

Jasmine 是一个 JavaScript 的测试框架，Angular 中会自动安装它，从生成的 package.json 可以看出，如图 9.3 所示。

```
"@types/jasmine": "~2.8.8",
"@types/jasminewd2": "~2.0.3",
"codelyzer": "~4.5.0",
"jasmine-core": "~2.99.1",
"jasmine-spec-reporter": "~4.2.1",
"karma": "~3.1.1",
"karma-chrome-launcher": "~2.2.0",
"karma-coverage-istanbul-reporter": "~2.0.1",
"karma-jasmine": "~1.1.2",
"karma-jasmine-html-reporter": "^0.2.2",
```

图 9.3　package.json 中的 Jasmine

Jasmine 并不依赖于任何其他的框架，如果需要在其他框架中使用它，可以直接使用以下指令单独安装它。

```
npm install --save-dev jasmine
```

通过分析源码，可以看到 Jasmine 给我们提供了丰富的匹配器（matcher），如图 9.4 所示。

```
index.d.ts
18  declare namespace jasmine {
19    interface Matchers<T> {
20      toBe(expected: any, expectationFailOutput?: any): Promise<void>;
21      toEqual(expected: any, expectationFailOutput?: any): Promise<void>;
22      toMatch(expected: string | RegExp | Promise<string | RegExp>, expectationFailOutput?: any): P
23      toBeDefined(expectationFailOutput?: any): Promise<void>;
24      toBeUndefined(expectationFailOutput?: any): Promise<void>;
25      toBeNull(expectationFailOutput?: any): Promise<void>;
26      toBeNaN(): Promise<void>;
27      toBeTruthy(expectationFailOutput?: any): Promise<void>;
28      toBeFalsy(expectationFailOutput?: any): Promise<void>;
29      toHaveBeenCalled(): Promise<void>;
30      toHaveBeenCalledWith(...params: any[]): Promise<void>;
31      toHaveBeenCalledTimes(expected: number | Promise<number>): Promise<void>;
32      toContain(expected: any, expectationFailOutput?: any): Promise<void>;
33      toBeLessThan(expected: number | Promise<number>, expectationFailOutput?: any): Promise<void>;
34      toBeLessThanOrEqual(expected: number | Promise<number>, expectationFailOutput?: any): Promise
35      toBeGreaterThan(expected: number | Promise<number>, expectationFailOutput?: any): Promise<voi
36      toBeGreaterThanOrEqual(expected: number | Promise<number>, expectationFailOutput?: any): Prom
37      toBeCloseTo(expected: number | Promise<number>, precision?: any, expectationFailOutput?: any)
38      toThrow(expected?: any): Promise<void>;
39      toThrowError(message?: string | RegExp | Promise<string | RegExp>): Promise<void>;
40      toThrowError(expected?: new (...args: any[]) => Error | Promise<new (...args: any[]) => Error
41    }
```

图 9.4　Jasmine 中的匹配器

接下来列举一下 Jasmine 内置的匹配器（matcher），参见表 9.1。

表 9.1　Jasmine 中的匹配器

匹配器	作用
toBe	结果是否相等，用===判断
toEqual	对象是否相等，如果是基本类型，则和 toBe 相同
toMatch	正则表达式
toBeDefined	是否不为 undefined
toBeUndefined	是否为 undefined
toBeNull	是否为 null
toBeNaN	是否为 NaN
toBeTruthy	是否为 true
toBeFalsy	是否为 false
toHaveBeenCalled	期望函数是否被调用
toHaveBeenCalledWith	期望函数是否被指定参数调用

（续表）

匹配器	作用
toHaveBeenCalledTimes	期望函数的调用次数
toContain	是否包含指定元素
toBeLessThan	是否小于指定值
toBeLessThanOrEqual	是否小于或等于指定值
toBeGreaterThan	是否大于指定值
toBeGreaterThanOrEqual	是否大于或等于指定值
toBeCloseTo	四舍五入后比较是否相等
toThrow	是否抛出错误
toThrowError	是否抛出指定错误

9.3.2 Karma

Karma 是一个基于 Node.js 的 JavaScript 测试执行过程管理工具（Test Runner）。

与 Jasmine 一样会被自动安装在项目中，它的主要作用是自动控制 Jasmine 编写单元测试。

对于 Karma，我们需要了解一下它的配置，它的主要配置文件就是 karma.conf.js，对它进行修改即可，当然使用它的默认配置也是没有问题的。

```
// Karma configuration file, see link for more information
// https://karma-runner.github.io/1.0/config/configuration-file.html

module.exports = function (config) {
  config.set({
    basePath: '',
    frameworks: ['jasmine', '@angular-devkit/build-angular'],
    plugins: [
      require('karma-jasmine'),
      require('karma-chrome-launcher'),
      require('karma-jasmine-html-reporter'),
      require('karma-coverage-istanbul-reporter'),
      require('@angular-devkit/build-angular/plugins/karma')
    ],
    client: {
      clearContext: false // leave Jasmine Spec Runner output visible in browser
    },
    coverageIstanbulReporter: {
      dir: require('path').join(__dirname, '../coverage'),
      reports: ['html', 'lcovonly', 'text-summary'],
      fixWebpackSourcePaths: true
    },
    reporters: ['progress', 'kjhtml'],
    port: 9876,
    colors: true,
    logLevel: config.LOG_INFO,
    autoWatch: true,
    browsers: ['Chrome'],
    singleRun: false
```

```
    });
};
```

【代码解析】它的配置参数比较多，比如 basePath 是相对位置，frameworks 是使用的库。我们看一个明显一些的例子，Karma 的 browsers 是指执行测试用的浏览器，默认为 Chrome，可以修改为其他浏览器，比如 Firefox、Safari、Opera 等等。这里的参数就不再一一列举了，我们这次就以默认的参数进行测试，如果读者需要了解其他参数，则可以查阅 Karma 的官方文档 http://karma-runner.github.io/latest/index.html。

9.3.3　实战练习：单元测试常用 API

本小节通过创建一个实战练习，来展示如何使用 Jasmine 和 Karma 来进行单元测试，并把常用的 API 都过一遍。

1. 服务的单元测试

在我们的日常开发中，经常会用到一些方便开发的工具类，里面会写上很多诸如验证手机号、验证邮箱等小工具。那么为了使这个工具类可以正常、准确地返回我们想要的结果，单元测试就显得很重要了。我们继续使用刚才新建的项目，输入以下指令新建一个服务。

```
ng g service service/utils
```

有一点需要注意，指令参数中加了 service/的话，就会在 app 文件夹下新建一个名为 service 的文件夹，这样就方便我们对 CLI 生成的文件进行管理了，最终的文件夹结构如图 9.5 所示。

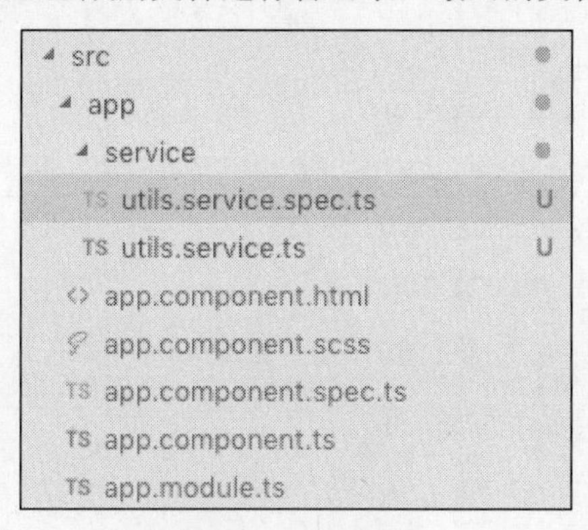

图 9.5　utils 服务

接下来在 utils.service.ts 中编写两个方法作为例子。

```
import { Injectable } from '@angular/core';

@Injectable({
  providedIn: 'root'
})
export class UtilsService {

  constructor() { }

  getUserName(): string {
```

```
    if (localStorage.getItem('username')) {
      return localStorage.getItem('username');
    } else {
      return '匿名用户';
    }
  }

  validateMobile(mobile: string): boolean {
    if
(!(/(^(13\\d|15[^4,\\D]|17[135678]|18\\d)\\d{8}|170[^346,\\D]\\d{7})$/.test(mo
bile))) {
      return true;
    } else {
      return false;
    }
  }
}
```

【代码解析】第一个方法 getUserName 是用来从 localStorage 获取用户名，若没有取到，则返回匿名用户。第二个方法则是使用正则表达式验证是否为手机号。接下来会对这两个方法进行单元测试。

打开 utils.service.spec.ts，可以看到系统创建服务时已经自动编写了一个单元测试了。不用管它，再编写两个单元测试，分别对上面的两个方法进行测试，代码如下：

```
import { TestBed } from '@angular/core/testing';

import { UtilsService } from './utils.service';

describe('UtilsService', () => {
  let utilsService: UtilsService;
  beforeEach(() => {
    TestBed.configureTestingModule({});
    utilsService = new UtilsService();
  });

  it('should be created', () => {
    const service: UtilsService = TestBed.get(UtilsService);
    expect(service).toBeTruthy();
  });

  it('成功获取用户名', () => {
    console.log(utilsService.getUserName());
    expect(utilsService.getUserName()).toBeDefined();
  });

  it('手机号是否合法', () => {
    console.log(utilsService.validateMobile('18622797149'));
    expect(utilsService.validateMobile('18622797149')).toBeTruthy();
  });
});
```

【代码解析】在成功获取用户名的单元测试中，测试了返回值是否不为 undefined 来判断是否有值，并且输出了返回值。在手机号是否合法的单元测试中，我们同样先输出了返回值，并用

toBeTruthy 来判断返回值是否为真，读者也可以试着将传入的手机号修改一下，来测试是否可以正常使用。数据类型则不需要担心，由于使用了 TypeScript 设置必须传 string 类型，其他类型是传不进去的。服务类单元测试的结果如图 9.6 所示。

```
LOG: '匿名用户'
Chrome 72.0.3626 (Mac OS X 10.14.2): Executed 4 of 6 SUCCESS (0 secs / 0.15 secs)
LOG: true
Chrome 72.0.3626 (Mac OS X 10.14.2): Executed 5 of 6 SUCCESS (0 secs / 0.151 secs)
Chrome 72.0.3626 (Mac OS X 10.14.2): Executed 6 of 6 SUCCESS (0.168 secs / 0.152 secs)
TOTAL: 6 SUCCESS
TOTAL: 6 SUCCESS
```

图 9.6　服务类单元测试结果

2. 组件的单元测试

接下来测试一下我们最常用到的组件。输入以下指令新建一个组件。

```
ng g component component/userInfo
```

接下来填写 user-info.component 组件的内容。输入以下代码：

```
// user-info.component.html
<nz-tag [nzColor]="'blue'">用户名：张三</nz-tag>
<button nz-button nzType="primary" (click)="changeStatus()">用户状态：
{{status}}</button>

// user-info.component.ts
import { Component, OnInit } from '@angular/core';

@Component({
  selector: 'app-user-info',
  templateUrl: './user-info.component.html',
  styleUrls: ['./user-info.component.scss']
})
export class UserInfoComponent implements OnInit {

  status = '停用';
  constructor() { }
  ngOnInit() { }

  changeStatus() {
    if (this.status == '停用') {
      this.status = '正常';
    } else {
      this.status = '停用';
    }
  }
}

// app.component.html
<h1>Welcome to ngtest!</h1>
<app-user-info></app-user-info>
```

【代码解析】在这个用户信息组件中，我们列出了用户名，并且增加了一个按钮控制用户的状态，单击后会在正常/停用两种状态切换。最后不要忘了在 app.component.html 中修改代码将组件显示出来。组件的显示结果如图 9.7 所示。

图 9.7　用户信息组件

这个时候我们会发现测试用例报了好几个错误，如图 9.8 所示。这些错误一共有两个原因：一是 user-info 组件没有在 app.component.spec.ts 导入。二是在 user-info 组件中用到了 NG-ZORRO，但是 user-info.component.spec.ts 和 app.component.spec.ts 没有导入相关的模块（Module）。

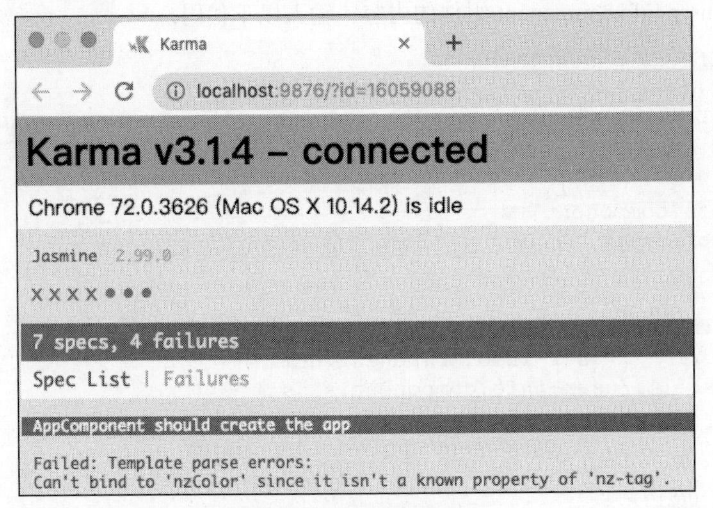

图 9.8　用户信息组件

这也就是我们为什么要在 user-info 组件中使用 nz-tag 这样一个属于 NG-ZORRO 的组件的原因。在我们日常开发中，很可能要使用到第三方的组件，这样在编写单元测试时，就会遇到这样的问题。在 app.component.spec.ts 输入以下代码导入对应的模块（Module）即可，需要注意的是，不只是 NG-ZORRO 的模块（Module）要导入，还有 FormsModule 等模块同样需要被导入，注意看测试报错的提示，它会给出缺少的项是什么的相应信息。

```
    import { TestBed, async } from '@angular/core/testing';
    import { AppComponent } from './app.component';
    import { NgZorroAntdModule } from 'ng-zorro-antd';
    import { FormsModule } from '@angular/forms';
    import { UserInfoComponent } from
'./component/user-info/user-info.component';
    import { BrowserAnimationsModule, NoopAnimationsModule } from
'@angular/platform-browser/animations';

    describe('AppComponent', () => {
```

```
beforeEach(async(() => {
  TestBed.configureTestingModule({
    imports: [BrowserAnimationsModule, NoopAnimationsModule, FormsModule,
NgZorroAntdModule],
    declarations: [
      AppComponent,
      UserInfoComponent
    ],
  }).compileComponents();
}));

// 以下省略...
});
});
```

【代码解析】其实理解起来比较简单。TestBed 在测试的时候，同样需要将依赖到的所有组件、模块（Module）都进行导入，其方式与我们平时的导入方式完全相同。user-info.component. spec.ts 文件中的解决方式完全相同，这里就不再赘述。

接下来就该编写单元测试了。打开 user-info.component.spec.ts 文件，开始编写单元测试。

```
import { async, ComponentFixture, TestBed } from '@angular/core/testing';

import { UserInfoComponent } from './user-info.component';
import { NgZorroAntdModule } from 'ng-zorro-antd';
import { FormsModule } from '@angular/forms';

describe('UserInfoComponent', () => {
  let component: UserInfoComponent;
  let fixture: ComponentFixture<UserInfoComponent>;

  beforeEach(async(() => {
    TestBed.configureTestingModule({
      imports: [FormsModule, NgZorroAntdModule],
      declarations: [UserInfoComponent]
    })
      .compileComponents();
  }));

  beforeEach(() => {
    fixture = TestBed.createComponent(UserInfoComponent);
    component = fixture.componentInstance;
    fixture.detectChanges();
  });

  it('should create', () => {
    expect(component).toBeTruthy();
  });

  it('使用开关控制用户状态', () => {
    component.changeStatus();
    expect(component.status).toBe('正常');
    component.changeStatus();
    expect(component.status).toBe('停用');
  });
```

```
});
```

【代码解析】上面的代码同样是生成的，与自动生成的服务测试用例不同，组件的测试用例帮我们新建了一个 component 的对象。我们同样可以使用它来调用组件内的方法。我们在“使用开关控制用户状态”这一测试中，首先调用了修改状态按钮，测试用户状态是否正常。之后再次改变测试是否停用。如果代码没有问题的话，8 个单元测试应该会全部通过。测试结果如图 9.9 所示。

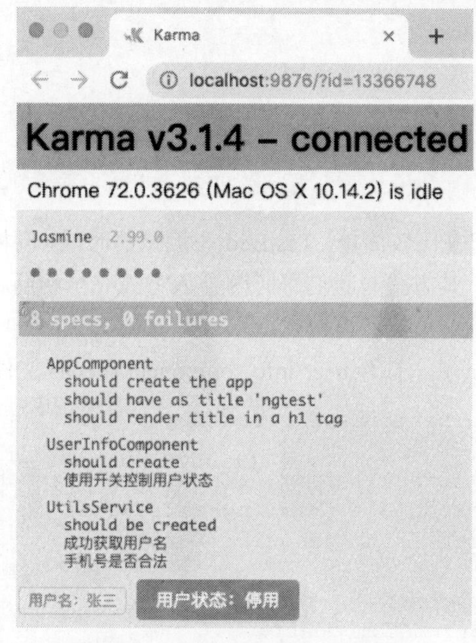

图 9.9 8 个单元测试全部通过

9.4 端对端测试

端对端测试（End to End Test）与前面的单元测试不同，端对端测试实际上是模仿人工操作来进行测试，比如单击“登录”按钮是否跳转到了我们预期的页面等等。在 Angular CLI 构建项目的过程中，同样会为端对端测试生成相关的工具与配置的相关文件，它的主要文件夹存放在 e2e 这个文件夹中。

9.4.1 Protractor

如果说 Jasmine 是一个 JavaScript 的测试框架，Protractor 就是 Angular 的专属测试框架了。它的测试方式是直接在真实的浏览器中运行，并以用户的身份进行交互操作。

9.4.2 实战练习：端对端测试常用 API

这次需要在新项目中创建一个登录组件，以登录业务作为测试点，编写一系列的端对端测试代码。

通过以下代码新建一个完整的项目。

```
ng new nge2e
ng add ng-zorro-antd
ng g component component/login
ng g component component/home
```

接下来需要把相关路由的代码编写一下，用于登录的跳转。

```
// app-routing.module.ts
import { NgModule } from '@angular/core';
import { Routes, RouterModule } from '@angular/router';
import { LoginComponent } from './login/login.component';
import { HomeComponent } from './home/home.component';

const routes: Routes = [
  { path: '', component: LoginComponent },
  { path: 'login', component: LoginComponent },
  { path: 'home', component: HomeComponent }
];

@NgModule({
  imports: [RouterModule.forRoot(routes)],
  exports: [RouterModule]
})
export class AppRoutingModule { }

// app.component.html
<router-outlet></router-outlet>

// app.module.ts
import { BrowserModule } from '@angular/platform-browser';
import { NgModule } from '@angular/core';

import { AppComponent } from './app.component';
import { NgZorroAntdModule, NZ_I18N, en_US } from 'ng-zorro-antd';
import { FormsModule } from '@angular/forms';
import { HttpClientModule } from '@angular/common/http';
import { NoopAnimationsModule } from '@angular/platform-browser/animations';
import { registerLocaleData } from '@angular/common';
import en from '@angular/common/locales/en';
import { LoginComponent } from './login/login.component';
import { HomeComponent } from './home/home.component';
import { AppRoutingModule } from './app-routing.module';

registerLocaleData(en);

@NgModule({
  declarations: [
    AppComponent,
    LoginComponent,
    HomeComponent
  ],
  imports: [
```

```
    BrowserModule,
    AppRoutingModule,
    NgZorroAntdModule,
    FormsModule,
    HttpClientModule,
    NoopAnimationsModule
  ],
  providers: [{ provide: NZ_I18N, useValue: en_US }],
  bootstrap: [AppComponent]
})
export class AppModule { }
```

然后是登录的具体代码，可以复用一下第 7 章的登录代码，将其稍微进行修改即可。

```
// login.component.html
<form (ngSubmit)="onSubmit()" #loginForm="ngForm">
  <div style="width: 300px;margin: 50px 50px">
    <input id="e2e-username" class="distance" type="text" nz-input
placeholder="请输入账号" required minlength="6" [(ngModel)]="username" name="name"
      #name="ngModel">
    <div *ngIf="name.errors?.minlength">
      账号必须大于 6 位
    </div>
    <input id="e2e-password" class="distance" type="password" nz-input
placeholder="请输入密码" required minlength="6" [(ngModel)]="passowrd"
      name="pwd" #pwd="ngModel">
    <div *ngIf="pwd.errors?.minlength">
      密码必须大于 6 位
    </div>
    <label class="distance" nz-checkbox [(ngModel)]="remember" name="check">
      <span>记住我</span>
    </label>
    <a class="distance" style="float:right">忘记密码</a>
    <button id="e2e-login-button" type="submit"
[disabled]="!loginForm.form.valid" class="distance" nz-button
[nzType]="'primary'" style="width: 300px">登录</button>
  </div>
</form>

// login.component.scss
.distance {
  margin-top: 20px;
}

// login.component.ts
import { Component, OnInit } from '@angular/core';
import { Router } from '@angular/router';

@Component({
  selector: 'app-login',
  templateUrl: './login.component.html',
  styleUrls: ['./login.component.scss']
})
export class LoginComponent implements OnInit {
```

```
    username = '';
    passowrd = '';
    remember = false;
    constructor(private router: Router) { }
    ngOnInit() {
    }

    onSubmit() {
      console.log(`账号：${this.username} 密码：${this.passowrd} 记住密码：
${this.remember}`);
      this.router.navigate(['/home']);
    }
  }

  // home.component.html
  <p>
    这是首页，登录成功。
  </p>
```

【代码解析】代码逻辑不是很复杂，如果输入了六位以上的账号和密码，单击"登录"按钮即可通过路由跳转到首页，其标签中的 id 是用来在端对端测试时使用的。程序执行的结果如图 9.10 所示。

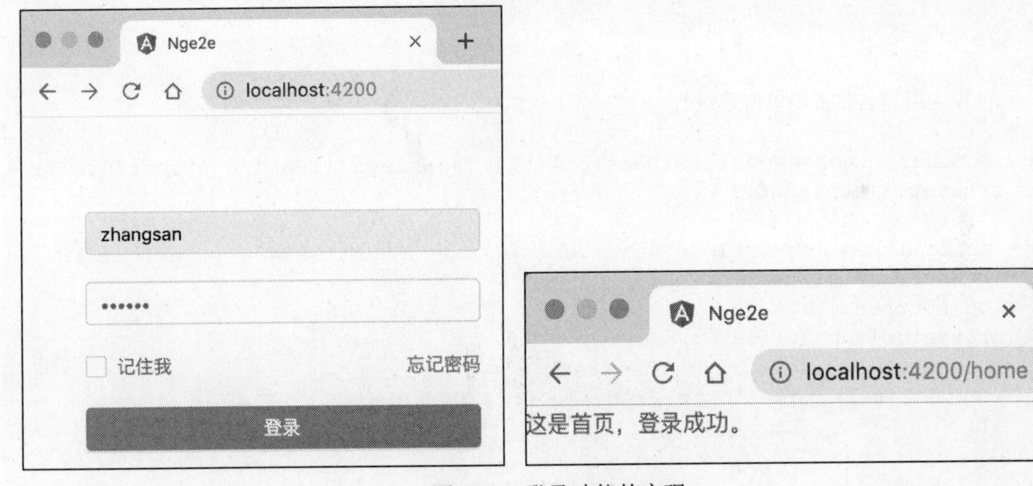

图 9.10　登录功能的实现

现在业务代码已经完成，接下来需要编写端对端测试代码对这项业务进行测试。将 e2e/src/app.e2e-spec.ts 文件修改为以下代码：

```
  import { AppPage } from './app.po';
  import { browser, by, element } from 'protractor';

  describe('workspace-project App', () => {
    let page: AppPage;

    beforeEach(() => {
      page = new AppPage();
      browser.get('');
    });
```

```
    // 用户名
    const username: any = element(by.id('e2e-username'));
    // 密码
    const password: any = element(by.id('e2e-password'));
    // 登录按钮
    const loginButton: any = element(by.id('e2e-login-button'));

    it('不输入用户名和密码单击登录', () => {
      loginButton.click();
      browser.getCurrentUrl().then(res => {
        expect(res).toBe('http://localhost:4200/');
      });
    });

    it('不输入密码单击登录', () => {

browser.actions().mouseMove(username).click().sendKeys('zhangsan').perform();
      browser.sleep(2000);
      loginButton.click();
      browser.sleep(2000);
      browser.getCurrentUrl().then(res => {
        expect(res).toBe('http://localhost:4200/');
      });
    });

    it('输入用户名和密码单击登录', () => {

browser.actions().mouseMove(username).click().sendKeys('zhangsan').perform();
      browser.sleep(2000);

browser.actions().mouseMove(password).click().sendKeys('123456').perform();
      browser.sleep(2000);
      loginButton.click();
      browser.sleep(2000);
      browser.getCurrentUrl().then(res => {
        expect(res).toBe('http://localhost:4200/home');
      });
    });
  });
```

【代码解析】我们一共编写了四个端对端的测试。分别是不输入用户名与密码单击登录、不输入密码单击登录、输入账号和密码单击登录。可以看出我们的逻辑基本是通过 element(by.id)来获取控件，之后进行输入或者单击操作，在最后使用 expect 来对比结果。其中 browser.sleep()主要是用来让操作慢一些，更容易看清浏览器的自动操作。

现在就可以使用 ng e2e 进行测试了，需要注意的一点是，这个指令会下载一个控制浏览器的工具，在命令行中能看到这条代码。

```
    I/downloader - curl
-o/Users/Lym/Desktop/MyProject/Angular-example/Section9/nge2e/node_modules/pro
tractor/node_modules/webdriver-manager/selenium/chromedriver_2.46.zip
https://chromedriver.storage.googleapis.com/2.46/chromedriver_mac64.zip
```

这个下载如果失败的话，可以到 http://chromedriver.storage.googleapis.com/index.html 下载，并将文件放入该文件夹中。

之后每次运行时都会自动更新，若不需要的话，则使用以下指令来执行。

```
ng e2e --webdriverUpdate=false
```

最后如果没有特殊情况，就可以看到 Chrome 自动启动，并且成功完成了这个端对端的测试。端到端的测试结果如图 9.11 所示。

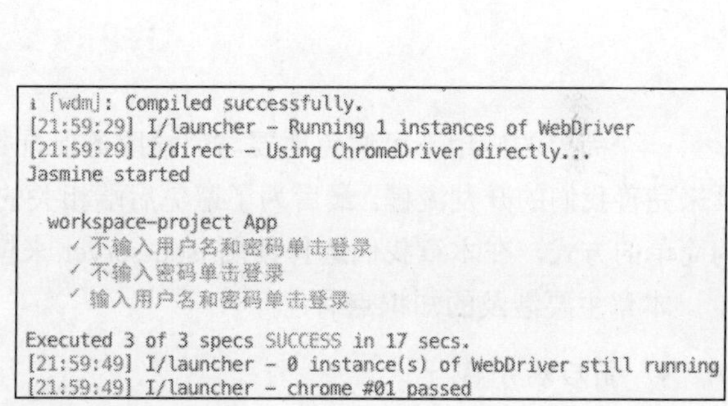

图 9.11 登录功能的端对端测试

9.5 小 结

本章首先讲解了测试的意义和一个简单的单元测试的例子，然后将 Angular 中的测试工具 Jasmine 和 Karma 的使用方式结合一个例子进行了讲解，最后通过一个对于登录业务的端对端测试的例子介绍了 Protractor 框架。

由于篇幅所限，所以测试功能无法讲得更全面、深入。如果读者对于这些测试框架感兴趣，可以搜索它们的官网，查阅官网提供的文档进行学习。下一章将搭建后台模拟环境，以便于我们在最后的实战章节使用。

第10章

后台模拟环境的搭建

一个完整的应用不可避免地要与后端服务保持数据交互，本章将介绍如何使用 Postman 等工具来完善我们的开发流程。最后为了避免后端相关的知识给读者的理解带来麻烦，我们尽量使用较为简单的方式，在本章我们选择使用 json-server 来模拟后端服务。

本章主要涉及的知识点有：

- 前后端分离
- Postman 的安装与使用
- json-server 的安装与使用
- 实战练习：使用 json-server 实现增删改查

10.1 前后端分离

如本章开头所述，除了一些工具类的单机应用外，大多数完整的应用离不开与后端的交互，所以这里简单谈一谈前后端为什么要分离。

以前的开发人员前端、后端都要做，导致职责划分不清晰，而且要学的知识面过于宽广，无法专精一个方向。后来慢慢就提出了前后端分离这个概念，让不同的开发人员各司其职。从图 10.1 可以看出，前端负责 View（视图）和 Controller（控制器）层，用于页面的展示等，而后端负责 Model（模型）层、业务处理和数据等。

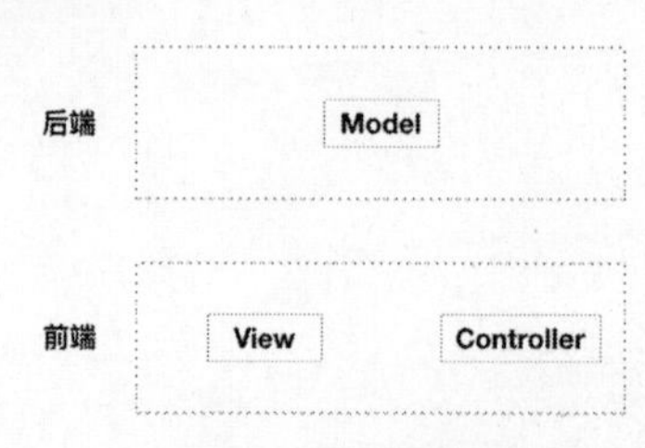

图 10.1　前后端分离

由于前后端的分离，初级前端工程师一般对后端知识接触得不多，所以这一章选择 json-server 模拟后端数据。

10.2　Postman 的安装与使用

使用 Angular 开发出的网页本质是一个单页面应用（Single Page Application）。这个应用与后端服务器的交互将主要通过调用后端服务器提供的 Restful 风格的 API 来实现。而测试验证后端服务器 API 有效性的需求，就催生了像 Postman 这样的浏览器插件工具在开发人员中的流行。

按照官方网站的宣传，Postman 是专门用于帮助开发人员快速开发 API 的工具。具体来说，Postman 允许用户发送任何类型的 HTTP 请求，包括 Restful API 使用到的 GET、POST、HEAD、PUT、DELETE 等，并且可以由开发人员任意定制参数和 HTTP 头（Headers）。此外，Postman 的输出是自动按照语法格式给出语法解析的结果，目前它支持的常见语法包括 HTML、JSON 和 XML。

10.2.1　Postman 的安装

安装 Postman 需要先登录到它的官方网站：https://www.getpostman.com，找到安装入口，如图 10.2 所示。

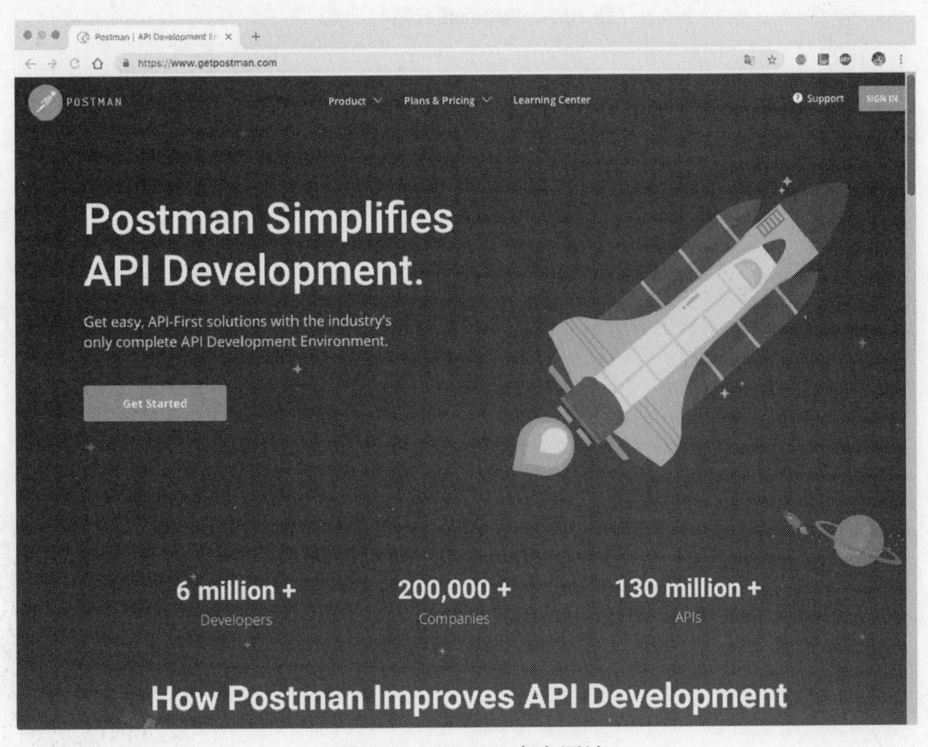

图 10.2　Postman 官方网站

随后直接单击 Get Started 按钮，该网站会自动根据用户开发环境所在的操作系统进入相应的下载页面，如图 10.3 所示。

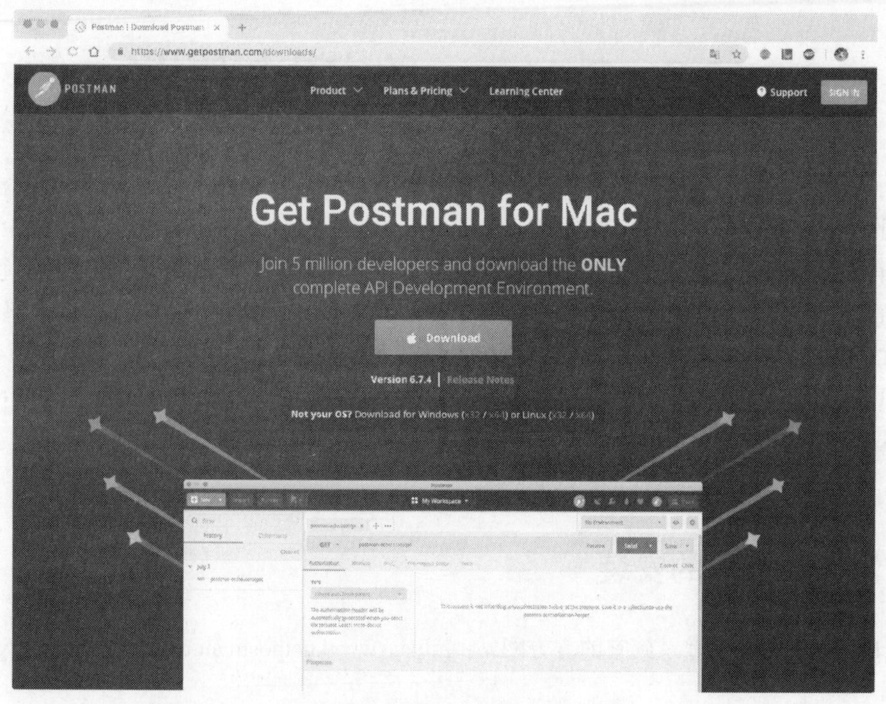

图 10.3 Postman 安装与启动

打开 Postman 后，可以根据界面上的元素直观地找到输入 HTTP 请求 URL 的输入框，如图 10.4 所示。

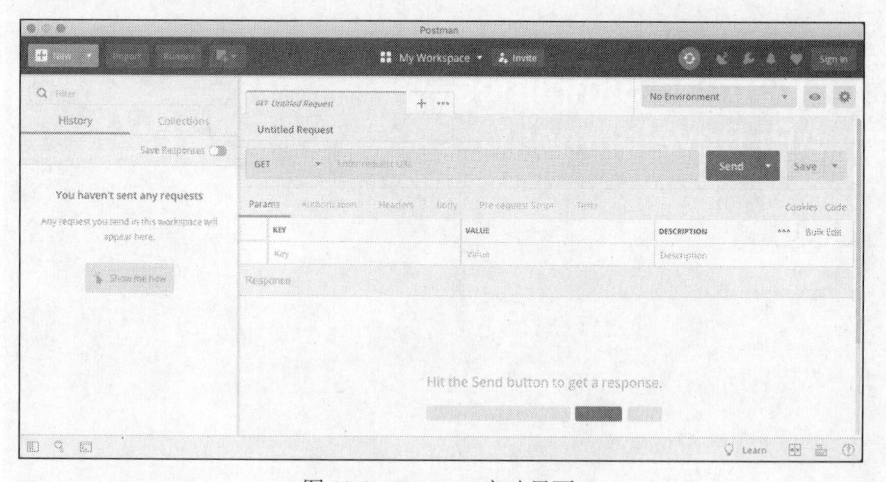

图 10.4 Postman 启动界面

10.2.2 Postman 的使用

作为简单示例，这里给读者推荐个 JSON 测试网站，我们就用它的 API 来进行测试：

http://jsonplaceholder.typicode.com/comments?postId=1，如图 10.5 所示。

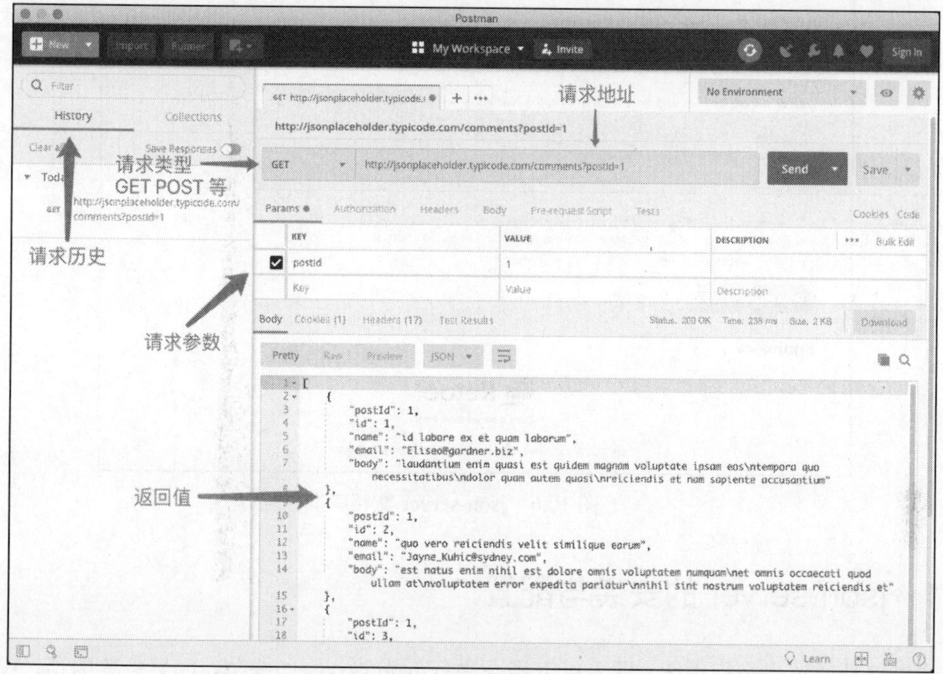

图 10.5　Postman 返回的 JSON 对象

需要经常使用的一些定制 HTTP 请求的配置选项，如 HTTP Method、HTTP 参数和验证方式请参见图 10.5 中的文字标注。从图中可以看到，该请求最后返回了来自 jsonplaceholder 的参数。

在确认 Postman 能正常工作后，下一节将安装 json-server 并使用它开发一个不连接数据库的简单数据维护 API，随后再使用 Postman 来测试这个 API。

10.3　json-server 的安装与使用

json-server 是一个开源的框架，可以在不写一句代码的情况下，实现 Rest API，是前端开发人员模拟后端服务的优秀工具之一。该框架在 GitHub 中的 Star 已经有 38000 余个，可以说是相当受欢迎，如图 10.6 所示。

如果在使用 json-server 的过程中，有什么问题或者建议，可以到 https://github.com/typicode/json-server 发起 Issue 和 Pull Request，为开源事业贡献自己的一份力量。下面开始讲解如何安装和使用 json-server。

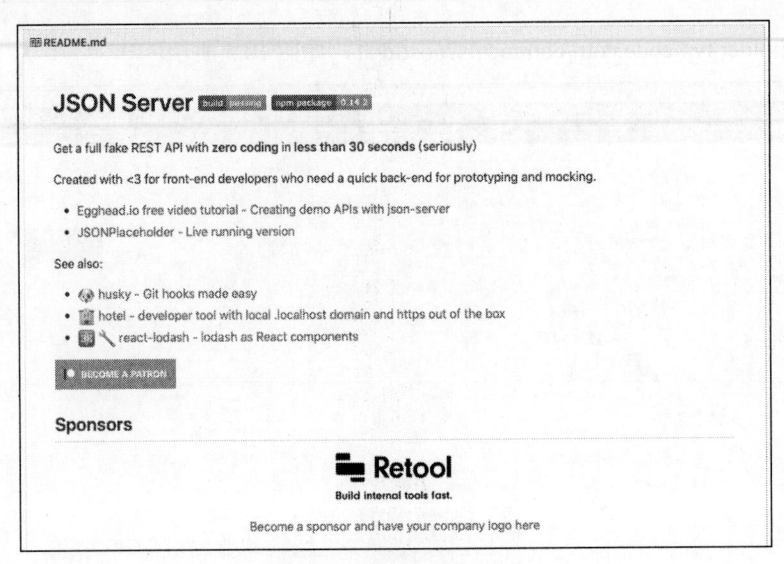

图 10.6　json-server 文档

10.3.1　json-server 的安装与配置

安装依然是通过命令行进行，输入以下指令进行 json-server 的安装。如果失败的话，Windows 用户使用管理员身份启动“命令提示符”应用程序，而 MAC 用户则需要在指令前加上 sudo。安装过程如图 10.7 所示。

```
npm install -g json-server
```

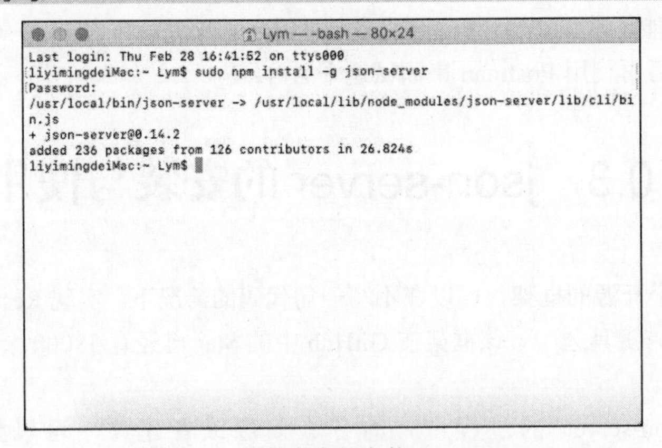

图 10.7　使用命令行安装 json-server

安装成功后，需要详细了解一下 json-server 的配置参数。作为一个模拟 Rest API 的工具，了解配置参数有助于我们更加高效地使用这个工具，否则只按照默认配置来使用的话，很可能会忽略掉一些便于使用的属性。

json-server 的参数使用方式很简单，在命令行按照以下格式输入参数即可。全配置参数如表 10.1 所示。

```
json-server [options] <source>
```

表 10.1　json-server 全配置参数

参数	简称	说明	默认值
--config	-c	指定配置文件	json-server.json
--port	-p	设置端口	3000
--host	-H	设置域	0.0.0.0
--watch	-w	是否监听	false
--routes	-r	指定自定义路由	
--middlewares	-m	指定中间件文件路径	
--static	-s	设置静态文件目录	
--read-only	--ro	是否只读（只用 GET）	false
--no-cors	--nc	是否禁用跨域	false
--no-gzip	--ng	是否禁用 GZIP 内容编码	false
--snapshots	-S	设置预览目录	.
--delay	-d	设置请求延迟时间	0
--id	-i	设置数据库 ID 属性	id
--foreignKeySuffix	--fks	设置外键后缀	
--quiet	-q	禁止输出日志	false
--help	-h	查看帮助信息	
--version	-v	查看版本号	

接下来我们对其中几个常用的配置项进行测试。比如我们一开始运行的时候，域名端口默认为 3000，如果这个端口被占用了，我们想换一个自定义的端口，输入以下指令即可实现。程序的执行结果如图 10.8 所示。

```
json-server --port 8100 data.json
```

图 10.8　使用配置参数替换端口

再举一个例子，在使用 json-server 的过程中，我们对源文件 data.json 进行修改后，它不会立即生效，需要重启才能生效。如果加上监听参数，它就会监听你的文件，如果内容出现变化，

json-server 会自动重新加载，这时马上就能请求到最新的接口内容。输入以下指令即可实现。

```
json-server --watch --port 8100 data.json
```

可以看到，这次输入的配置参数是直接在--port 的前面增加了--watch。从这里可以看出，如果想一次加载多个配置参数，只需要增加一个空格直接配置即可。

10.3.2 第一个 json-server 程序

【示例 10-1】首先还是从快速实现一个最简单的程序入手，之后再一步步讲解它是如何实现的。在命令行执行以下指令。

```
json-server data.json
```

如果在命令行中输出了以下内容，就说明 json-server 运行成功了。

```
\{^_^}/ hi!

Loading data.json
Done

Resources
http://localhost:3000/posts
http://localhost:3000/comments
http://localhost:3000/profile

Home
http://localhost:3000

Type s + enter at any time to create a
snapshot of the database
Watching...
```

打开 http://localhost:3000，可以看到一个引导页，如图 10.9 所示。

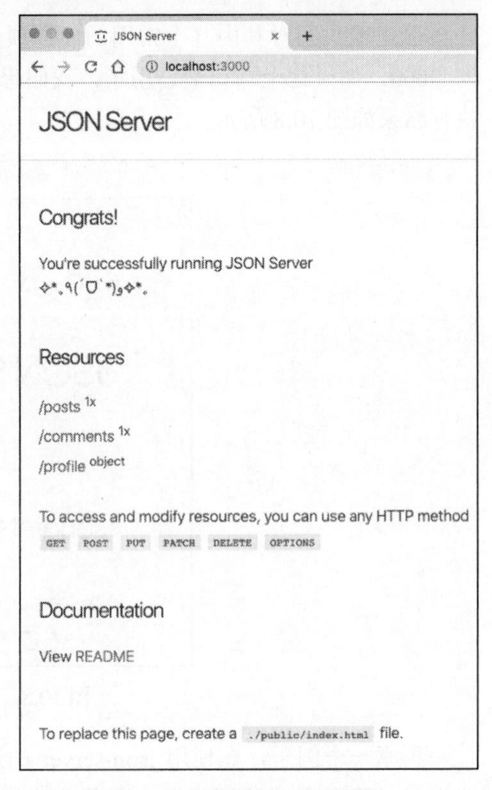

图 10.9 json-server 引导页

首先分析一下这个页面的内容。顶部是 json-server 的名称 JSON Server，下面一段是祝贺成功运行。Resources 则是我们需要关注的重点，单击上面的/posts、/comments、/profile 都可以看到对应的数据。1x 说明这个数据是数组类型，里面有一个元素。object 则说明是一个对象类型。Documentation 部分放上了官方说明文档的地址。底部还提示说可以创建一个 index.html 来替换这个页面。

接下来使用 Postman 来测试一下它生成的接口是否可以正常使用。打开 Postman 并输入 http://localhost:3000/posts，单击 Send 按钮查看结果，如图 10.10 所示。

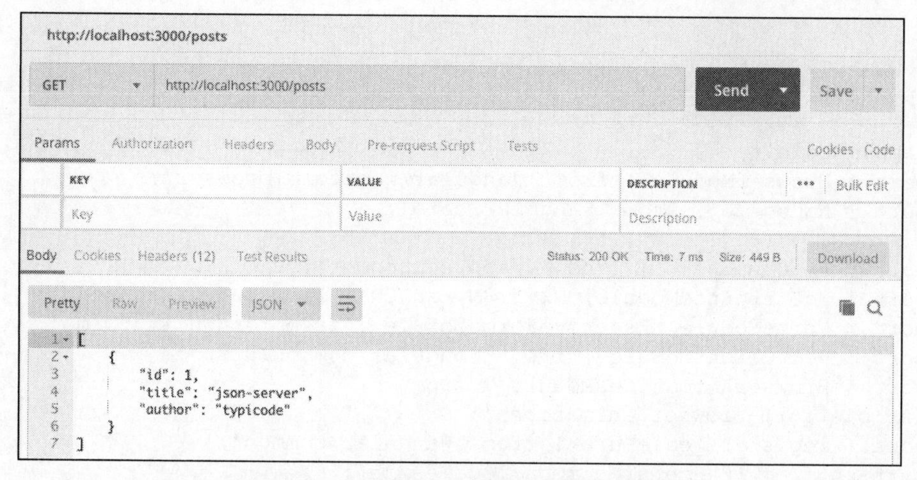

图 10.10　使用 Postman 请求 json-server 接口

可以看到已经能通过 Postman 调用到接口了，这就说明使用 Angular 也可以调用这个接口了。但是它只能做查询，还不能完全支持我们在实战项目中需要的完整功能，至少要实现增删改查功能。下一节将通过一个实战练习，全面掌握增删改查的方法。

10.4　实战练习：使用 json-server 实现增删改查

相较于长篇幅的文档，使用实例的方式更容易掌握知识点，对于工具类的学习更是如此。这一小节将使用 json-server 实现增删改查，并和 Angular 应用进行调用，构建出一个小而全的应用。

10.4.1　项目的创建与配置

首先还是要先创建项目。输入以下指令新建一个 Angular 项目，导入 NG-ZORRO 并新建一个 json-server 文件。

```
ng new json-server-test
cd json-server-test
ng add ng-zorro-antd
json-server --watch data.json
```

需要注意的是，对于 json-server 是不可以通过 package.json 来进行设置的，让 Angular 和 json-server 同时启动。如果设置成以下指令，后面的指令将永远无法运行，因为&&是要等前一条指令执行完毕并且没有出错（返回 0）才会执行后一条指令。

```
ng serve && json-server --watch data.json6
```

接下来，新建一个表格组件和一个弹出式窗口组件，并将它写在 app.component.html 中。同样地，在 app.module.ts 中，我们也要检查是否将组件、HTTPModule、表单等都已注册完毕。

```
ng g component table
ng g component editModal
```

```
// app.component.html
<app-table></app-table>

// app.module.ts
import { BrowserModule } from '@angular/platform-browser';
import { NgModule } from '@angular/core';

import { AppComponent } from './app.component';
import { NgZorroAntdModule, NZ_I18N, zh_CN } from 'ng-zorro-antd';
import { FormsModule, ReactiveFormsModule } from '@angular/forms';
import { HttpClientModule } from '@angular/common/http';
import { BrowserAnimationsModule } from
'@angular/platform-browser/animations';
import { registerLocaleData } from '@angular/common';
import zh from '@angular/common/locales/zh';
import { TableComponent } from './table/table.component';
import { EditModalComponent } from './edit-modal/edit-modal.component';

registerLocaleData(zh);

@NgModule({
  declarations: [
    AppComponent,
    TableComponent,
    EditModalComponent
  ],
  imports: [
    BrowserModule,
    NgZorroAntdModule,
    FormsModule,
    ReactiveFormsModule,
    HttpClientModule,
    BrowserAnimationsModule
  ],
  providers: [{ provide: NZ_I18N, useValue: zh_CN }],
  bootstrap: [AppComponent]
})
export class AppModule { }
```

10.4.2 数据的查询与删除

为了便于展示增删改查的功能，我们将构建一个表格来直观地展示数据的操作。首先在 table 组件中搭建一个表格。

```
<button (click)="addPeople()" nz-button nzType="primary" style="margin: 20px;
width: 100px;">添加</button>

<nz-table style="padding: 20px" #myTable [nzData]="listData">
  <thead>
    <tr>
      <th>姓名</th>
      <th>年龄</th>
```

```
      <th>住址</th>
      <th>操作</th>
    </tr>
  </thead>
  <tbody>
    <tr *ngFor="let data of myTable.data">
      <td>{{data.name}}</td>
      <td>{{data.age}}</td>
      <td>{{data.address}}</td>
      <td>
        <a (click)="edit(data)">编辑</a>
        <nz-divider nzType="vertical"></nz-divider>
        <a (click)="delete(data)" style="color:red">删除</a>
      </td>
    </tr>
  </tbody>
</nz-table>
```

【代码解析】在这个页面中，顶部放了一个添加按钮。并使用 nz-table 建立了一个表格，分别显示姓名、年龄、住址和操作，操作功能中存放编辑与删除两个功能。

接下来把 TS 中的代码补充上去，这个表格就算完成了。

```
import { Component, OnInit } from '@angular/core';
import { HttpClient } from '@angular/common/http';
import { NzModalRef, NzModalService } from 'ng-zorro-antd';

@Component({
  selector: 'app-table',
  templateUrl: './table.component.html',
  styleUrls: ['./table.component.css']
})
export class TableComponent implements OnInit {

  // 列表数据
  listData = [];
  // 对话框
  confirmModal: NzModalRef;

  constructor(private http: HttpClient, private modal: NzModalService) { }

  ngOnInit() {
    this.getListData();
  }

  // 获取列表数据
  getListData() {
    this.http.get('http://localhost:3000/users').subscribe((res:
Array<Object>) => {
      this.listData = res;
    });
  }
  // 新增
  addPeople() {
  }
```

```
    // 编辑
    edit(data: Object) {
    }

    // 删除
    delete(data: Object) {
      this.confirmModal = this.modal.confirm({
        nzTitle: '提示',
        nzContent: '确定要删除吗？',
        nzOkText: '确定',
        nzCancelText: '取消',
        nzOnOk: () => {
          this.http.delete('http://localhost:3000/users/' +
data['id']).subscribe((res) => {
            this.getListData();
          });
        }
      });
    }
  }
```

【代码解析】页面中通过 getListData 方法获取列表数据，在 delete 中通过弹出式窗口提示调用接口删除数据，最后再重新获取一次列表。需要注意的是，查询方法需要使用 GET 请求，删除则使用 DELETE 请求。细心的读者会发现，新增和编辑的方法还是空着的。因为要给新增、编辑页单独制作一个弹出窗口，所以这个放到最后再来实现。

接下来在 data.json 中创造一些模拟数据，这个程序运行时就可以看到列表的内容了，如图 10.11 和图 10.12 所示。

```
  {
    "users": [
      {
        "id": 1,
        "name": "张三",
        "age": 18,
        "address": "北京市朝阳区"
      },
      {
        "id": 2,
        "name": "李四",
        "age": 20,
        "address": "天津市西青区"
      }
    ]
  }
```

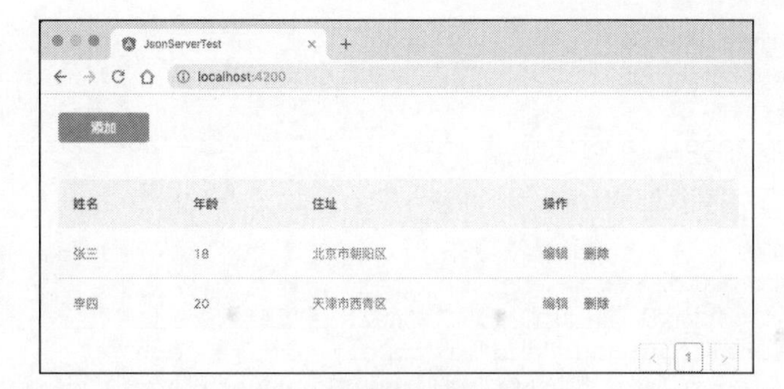

图 10.11　使用接口查询数据

图 10.12　使用接口删除数据

10.4.3　数据的新增与编辑

在上一节中，我们知道了查询和删除的 HTTP 分别为 GET 和 DELETE，那么新增和编辑就不难猜了，它们两个的方法分别是 POST 和 PUT，增删改查正好对应了最常用的四个请求方法。

其实在日常开发中，不一定是这四个方法分别对应增删改查。由于这个工具只是操作本地的 json 文件，为了方便和易于使用起见，所以固定设置成这样。接下来编写弹出式窗口对应的代码，制作新增与编辑功能。

```
    // edit-modal.component.html
    <nz-modal [(nzVisible)]="isVisible" nzTitle="新增人员" nzOkText="确定"
nzCancelText="取消" (nzOnCancel)="handleCancel()"
      (nzOnOk)="handleOk()">
      <form nz-form [formGroup]="validateForm" (ngSubmit)="submitForm()">
        <nz-form-item>
          <nz-form-label [nzSm]="6" [nzXs]="24" nzFor="name" nzRequired>
            <span>
              姓名
            </span>
          </nz-form-label>
          <nz-form-control [nzSm]="14" [nzXs]="24">
            <input nz-input id="name" formControlName="name">
            <nz-form-explain *ngIf="validateForm.get('name').dirty &&
validateForm.get('name').errors">请输入姓名!
            </nz-form-explain>
```

```
      </nz-form-control>
    </nz-form-item>
    <nz-form-item>
      <nz-form-label [nzSm]="6" [nzXs]="24" nzFor="age" nzRequired>
        <span>
          年龄
        </span>
      </nz-form-label>
      <nz-form-control [nzSm]="14" [nzXs]="24">
        <input nz-input id="age" formControlName="age">
        <nz-form-explain *ngIf="validateForm.get('age').dirty &&
validateForm.get('age').errors">请输入年龄!
        </nz-form-explain>
      </nz-form-control>
    </nz-form-item>
    <nz-form-item>
      <nz-form-label [nzSm]="6" [nzXs]="24" nzFor="address" nzRequired>
        <span>
          住址
        </span>
      </nz-form-label>
      <nz-form-control [nzSm]="14" [nzXs]="24">
        <input nz-input id="address" formControlName="address">
        <nz-form-explain *ngIf="validateForm.get('address').dirty &&
validateForm.get('address').errors">请输入住址!
        </nz-form-explain>
      </nz-form-control>
    </nz-form-item>
  </form>
</nz-modal>
```

【代码解析】这个页面最外层使用了 nz-modal 实现弹出式窗口，通过 isVisible 参数控制显示与隐藏，表单内部则使用模板驱动型表单进行数据校验。

接下来编写 TS 文件中的代码。

```
import { Component, OnInit, Input, Output, EventEmitter } from '@angular/core';
import { FormBuilder, FormGroup, Validators } from '@angular/forms';
import { HttpClient } from '@angular/common/http';

@Component({
  selector: 'app-edit-modal',
  templateUrl: './edit-modal.component.html',
  styleUrls: ['./edit-modal.component.css']
})
export class EditModalComponent implements OnInit {

  @Input()
  data: Object = {};

  @Input()
  isVisible = false; // 是否显示模态窗
  @Output()
  isVisibleChange = new EventEmitter(); // dialog 显示状态改变事件
  @Output()
```

```
  clickEvent = new EventEmitter();

  isEdit: boolean = false;
  validateForm: FormGroup;

  constructor(private fb: FormBuilder, private http: HttpClient) {
   this.validateForm = this.fb.group({
     name: [null, [Validators.required]],
     age: [null, [Validators.required]],
     address: [null, [Validators.required]]
   });
  }

  ngOnInit() { }

  ngOnChanges() {
    if (this.data['id']) {
      this.isEdit = true;
      this.validateForm.setValue({
        name: this.data['name'],
        age: this.data['age'],
        address: this.data['address']
      });
    } else {
      this.isEdit = false;
      this.validateForm.setValue({
        name: '',
        age: '',
        address: ''
      });
    }
  }

  handleOk(): void {
    this.submitForm();
  }

  handleCancel(): void {
    this.isVisibleChange.emit(false);
  }

  submitForm(): void {
    let params = {};
    for (const i in this.validateForm.controls) {
      this.validateForm.controls[i].markAsDirty();
      this.validateForm.controls[i].updateValueAndValidity();
      if (!(this.validateForm.controls[i].status == 'VALID') &&
this.validateForm.controls[i].status !== 'DISABLED') {
        return;
      }
      if (this.validateForm.controls[i] &&
this.validateForm.controls[i].value) {
        params[i] = this.validateForm.controls[i].value;
      } else {
        params[i] = '';
```

```
        }
      }
      if (this.isEdit) {
        this.http.put('http://localhost:3000/users/' + this.data['id'],
params).subscribe((res) => {
          this.clickEvent.emit();
        });
      } else {
        this.http.post('http://localhost:3000/users', params).subscribe((res)
=> {
          this.clickEvent.emit();
        });
      }
      this.isVisibleChange.emit(false);
    }
  }
```

【代码解析】这部分的代码比较长，我们从头开始一点点分析。首先页面导入了 HttpClient 与表单相关的依赖。传入参数包括：控制显示或隐藏的 isVisible，data 参数则是编辑时使用的，所以新增时不需要传值。之后在构造函数中初始化了 FormGroup，并在每次进入页面时使用 ngOnChanges 刷新里面的内容，否则可能会呈现旧数据。在完成新建或编辑后，会通过 clickEvent 通知父组件。

最后将 table 组件调整一下，就可以实现增删改查了。

```
    // table.component.html
    ...

    <app-edit-modal [data]="editData" [(isVisible)]="modalIsVisible"
(clickEvent)="clickEvent()">
    </app-edit-modal>

    // table.component.ts
    import { Component, OnInit } from '@angular/core';
    import { HttpClient } from '@angular/common/http';
    import { NzModalRef, NzModalService } from 'ng-zorro-antd';

    @Component({
      selector: 'app-table',
      templateUrl: './table.component.html',
      styleUrls: ['./table.component.css']
    })
    export class TableComponent implements OnInit {

      // 列表数据
      listData = [];
      // 编辑数据
      editData = {};
      // 模态窗显示
      modalIsVisible: boolean = false;
      // 对话框
      confirmModal: NzModalRef;
      constructor(private http: HttpClient, private modal: NzModalService) { }
      ngOnInit() {
```

```
      this.getListData();
    }

    // 获取列表数据
    getListData() {
      this.http.get('http://localhost:3000/users').subscribe((res: 
Array<Object>) => {
        this.listData = res;
      });
    }

    // 新增
    addPeople() {
      this.editData = {};
      this.modalIsVisible = true;
    }

    // 编辑
    edit(data: Object) {
      this.editData = data;
      this.modalIsVisible = true;
    }

    // 删除
    delete(data: Object) {
      this.confirmModal = this.modal.confirm({
        nzTitle: '提示',
        nzContent: '确定要删除吗？',
        nzOkText: '确定',
        nzCancelText: '取消',
        nzOnOk: () => {
          this.http.delete('http://localhost:3000/users/' + 
data['id']).subscribe((res) => {
            this.getListData();
          });
        }
      });
    }

    // 新增、编辑完成的回调
    clickEvent() {
      this.getListData();
    }
  }
```

【代码解析】代码变化不是很大，增加了编辑数据与控制模态窗显示的参数，在新增和编辑时将模态窗弹出，完成回调时再次使用 getListData 方法刷新页面即可。程序的执行结果如图 10.13 所示。

图 10.13 使用接口新增/编辑数据

在新增/编辑成功后，同样可以看到 data.json 文件的变动。

10.5 小 结

本章中从前后端分离开始讲起，重点介绍了两种实用的开发工具——Postman 与 json-server，以便为后面的使用做准备。在最后一个小节，我们通过一个人员增删改查的实战项目，帮助读者熟悉 json-server 的使用。本章作为一个过渡，内容不是很多，但是十分重要。如果在实战章节中，对于模拟后台和 Postman 的使用存在问题，可以回过头来复习一下本章相应的内容。

第 11 章

项目实战：待办列表

经过了前面各章的学习，我们终于来到了本书的实战部分。本书一共安排的两个实战项目，一个是偏向于复习组件、指令、TypeScript 的本地待办列表项目，另一个是更加完整的商城项目。本章的实战项目是待办列表，也就是我们常说的 TodoList。将完成目标时需要采取的每个步骤切割成一个个子目标，并列入待办事项列表中，是对达成任务极有助益的方法。

TodoList 如果作为个人使用，完全可以将数据存储在本地，而不需要后台服务。所以我们将项目分为两个部分，首先使用 LocalStorage 进行数据存储。待项目实现后，再使用 json-server 替换为带网络请求的应用。

本章主要涉及的知识点有：

- Angular 组件与指令
- TypeScript
- json-server
- 网络请求

11.1 待办列表设计

对于这个实战项目的设计，我们引用一个最近比较火的 MVP（Minimum Viable Product，最小化可行产品）概念来进行设计，它最早由埃里克·莱斯提出，刊载于哈弗商业评论。从图 11.1 可以大概分析一下 MVP 的基本理念。

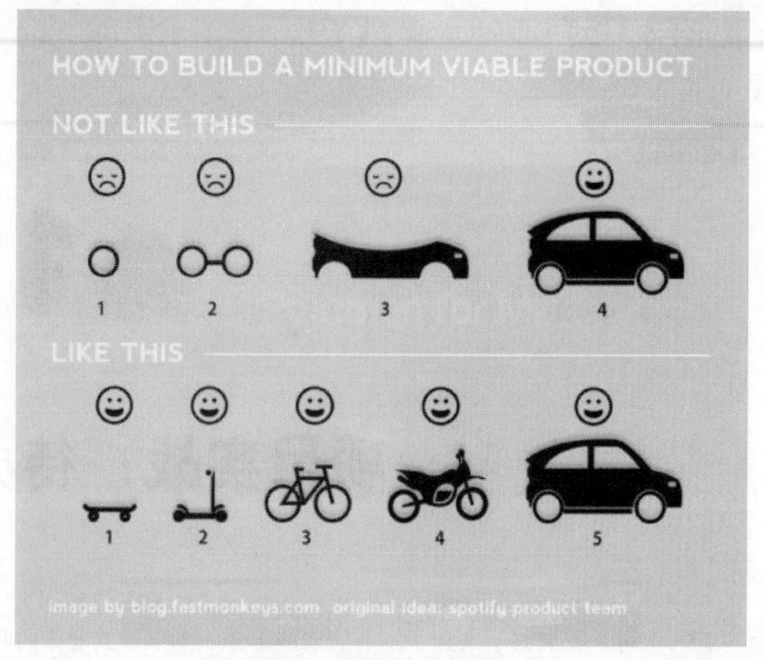

图 11.1 如何实现 MVP

可以看出，按照图 11.1 中第一种的方式（图中上半部分）所设计的，并不是一个可用的产品。而第二种设计方式的含义是，在产品处于雏形状态即有效可用。我们所要做的 TodoList，肯定无法一下子做出一个大而全的成品，但是至少要保证像 MVP 一样，达到最小可用的状态。

一个完整、可用的应用，首先肯定离不开增删改查，待办列表也是如此。所以我们的设计应该包括，增：新增待办，删：删除待办，改：编辑待办，查：查询（展示）待办，如图 11.2 所示。

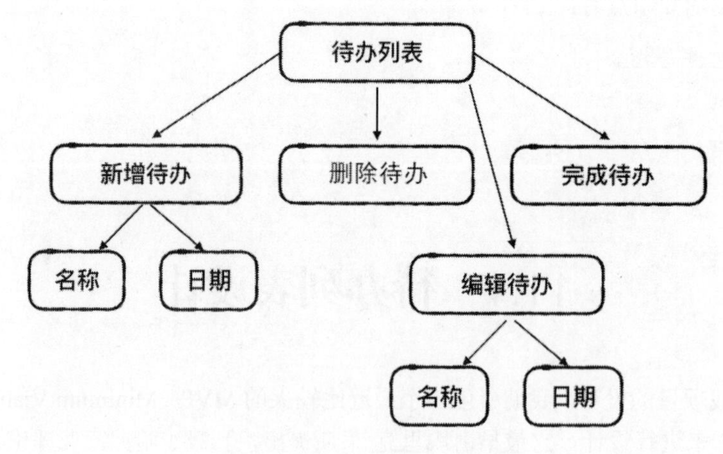

图 11.2 待办列表设计结构图

现在我们对要做一个什么样的应用有了一个大概的了解，下面继续学习待办列表项目的实现细节吧。

11.2 待办列表的创建

在上一节中，我们已经对项目做了需求分析与设计，相信读者已经有了整体印象。在本节中，我们会对项目需要用到的技术进行选型，并且完成目录结构、资源的搭建。好的开头是成功的一半，在开始编写代码之前，一定要规划好项目框架。

11.2.1 CLI 版本与 UI 样式库

首先我们来列举一下 CLI 的版本，如果有读者在运行项目的过程中出错了，有可能是与笔者的版本有差异。

```
Angular CLI: 7.1.0
Node: 10.11.0
NPM: 6.4.1
@angular-devkit/architect    0.11.0
@angular-devkit/core          7.1.0
@angular-devkit/schematics   7.1.0
@schematics/angular          7.1.0
@schematics/update            0.11.0
rxjs                         6.3.3
typescript                    3.1.6
```

在 UI 样式库中，我们依然和前面的章节一样，选择使用 NG-ZORRO 进行开发，版本号如下。

```
ng-zorro-antd: 7.0.0
```

11.2.2 项目的创建

选择好了所用的技术，我们就开始创建项目吧，输入以下指令创建 TodoList 项目。

```
ng new todolist
cd todolist
ng add ng-zorro-antd
```

之后根据模块的划分，将需要用到的组件一一搭建起来。

```
ng g component components/dashboard
ng g component components/edit-modal
ng g component components/item
```

简单介绍这三个组件的用途。

- dashboard：待办列表整体展示的容器。
- editModal：待办的添加、编辑的弹出式窗口。
- item：将待办选项组件化，方便复用。

接下来要确保 app.module.ts 已经注册了项目所需要的组件。之后打开 app.component.html，将里面的内容清空，并改为展示 dashboard 组件。

```
// app.module.ts
import { BrowserModule } from '@angular/platform-browser';
import { NgModule } from '@angular/core';

import { AppComponent } from './app.component';
import { NgZorroAntdModule, NZ_I18N, zh_CN } from 'ng-zorro-antd';
import { FormsModule } from '@angular/forms';
import { HttpClientModule } from '@angular/common/http';
import { NoopAnimationsModule } from '@angular/platform-browser/animations';
import { registerLocaleData } from '@angular/common';
import cn from '@angular/common/locales/zh';
import { DashboardComponent } from
'./components/dashboard/dashboard.component';
import { ItemComponent } from './components/item/item.component';
import { EditModalComponent } from
'./components/edit-modal/edit-modal.component';

registerLocaleData(cn);

@NgModule({
  declarations: [
    AppComponent,
    DashboardComponent,
    ItemComponent,
    EditModalComponent
  ],
  imports: [
    BrowserModule,
    NgZorroAntdModule,
    FormsModule,
    HttpClientModule,
    NoopAnimationsModule
  ],
  providers: [{ provide: NZ_I18N, useValue: zh_CN }],
  bootstrap: [AppComponent]
})
export class AppModule { }

// app.component.html
<app-dashboard></app-dashboard>
```

至此我们的框架搭建得差不多了，下一小节将正式开始开发。项目主要目录列表如图 11.3 所示，读者可以对照一下目录文件是否有缺漏。

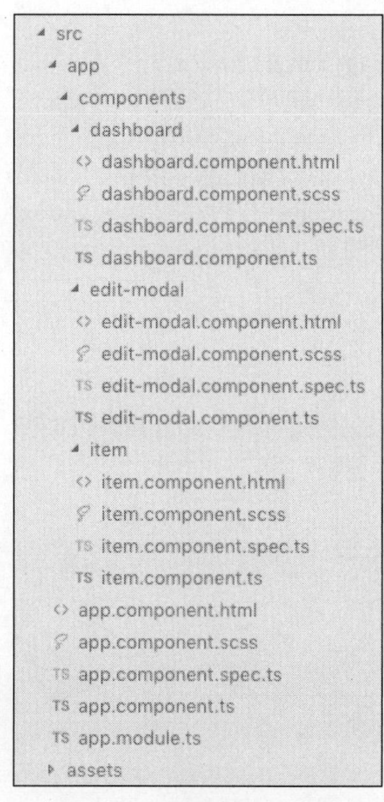

图 11.3　待办列表项目目录

11.3　待办列表的开发

现在项目的整体架构已经设计好了，但是对于开发的顺序还需要讲一下，不能东写一段西写一段，有规划的开发可以提高工作效率。从这三个组件来看，首先应该从 dashboard 入手，将页面构建起来，之后将待办项从中抽出，放到 item 组件中。最后再把添加、编辑的输入框提取出来放到 editModal 组件中。

11.3.1　主面板组件的开发

在主面板中，首先要把整个页面搭建起来。在制作弹出式窗口组件前，先简单做一个 input 输入框和一个按钮用于添加待办项。在 dashboard.component.html 和 dashboard.component.scss 输入以下代码：

```
    // dashboard.component.html
    <div class="dashboard">
      <h2>ToDoList</h2>
      <input nz-input [(ngModel)]="todoTitle" placeholder="添加 ToDo"
class="todo-input">
      <button nz-button nzType="primary" style="margin-left: 15px"
```

```
(click)="addTodo()">添加</button>
    <h2>正在进行 {{doingArray.length}}</h2>
    <div *ngFor="let item of doingArray;let i = index" style="margin-top: 15px">
      <app-item></app-item>
    </div>
    <h2>已经完成 {{doneArray.length}}</h2>
    <div *ngFor="let item of doneArray;let i = index" style="margin-top: 15px">
      <app-item></app-item>
    </div>
  </div>

  // dashboard.component.scss
  .dashboard {
    text-align: center;
    color: #2c3e50;
    margin-top: 60px;
  }

  .todo-input {
    margin-left: 15px;
    width: 300px;
  }

  .item {
    margin: 0px auto;
  }
```

【代码解析】这个页面可以分解为三个部分：顶部的输入框和添加按钮，中部的正在进行项和底部的已经完成项。我们对正在进行和已经完成的待办项使用数组的长度来展示选项的数量。最后通过 ngFor 遍历 doingArray、doneArray 来显示 item 组件。

接下来把 HTML 文件中需要用到的方法实现一下，这些方法可以说是业务逻辑。打开 dashboard.component.ts 输入以下代码：

```
  import { Component, OnInit } from '@angular/core';
  import { NzMessageService } from 'ng-zorro-antd';

  @Component({
    selector: 'app-dashboard',
    templateUrl: './dashboard.component.html',
    styleUrls: ['./dashboard.component.scss']
  })
  export class DashboardComponent implements OnInit {

    // 添加 Todo 标题
    private todoTitle = '';
    // 数据数组
    private dataArray: Array<Object> = [];
    // 待办数组
    private doingArray: Array<Object> = [];
    // 完成数组
    private doneArray: Array<Object> = [];
    constructor(private message: NzMessageService) { }
```

```
  // 生命周期
  ngOnInit() {
    this.getTodoList();
  }

  // 获取列表数据
  private getTodoList(): void {
    const dataString: string = localStorage.getItem('todo-list');
    if (dataString != null) {
      this.dataArray = JSON.parse(dataString);
      for (let i = 0; i < this.dataArray.length; i++) {
        const element: any = this.dataArray[i];
        if (element.done === true) {
          this.doneArray.push(element);
        } else {
          this.doingArray.push(element);
        }
      }
    }
  }

  // 添加 Todo
  private addTodo(): void {
    if (this.todoTitle === '') {
      this.message.info('请输入标题');
    } else {
      const item = {
        done: false,
        title: this.todoTitle
      };
      this.doingArray.push(item);
      this.dataArray.push(item);
      this.todoTitle = '';
      localStorage.setItem('todo-list', JSON.stringify(this.dataArray));
    }
  }
}
```

【代码解析】这一页的代码比较多，我们从变量开始分析。

- todoTitle：与 input 双向绑定，用于保存待办的标题和验证用户是否输入了完整的标题。
- dataArray：本地数据的数组，本地数据保存在 localStorage 中。
- doingArray：待办数组，由 dataArray 筛选而来。
- doneArray：已完成数组，由 dataArray 筛选而来。

介绍了声明的变量后，下面就来谈一谈方法。首先构造函数导入了 NG-ZORRO 提供的一个弹出式窗口，用于后面的未输入标题的提示。在生命周期中，我们调用了获取列表数据的方法，在列表数据方法中，我们通过 done 这个属性来确定待办项是否已经完成。最后 addTodo 方法用于添加待办事项，成功调用后，会往本地数组与 localStorage 同时存入一份数据。

下面来验收一下阶段性的成果。分别不输入标题和输入标题单击确认，程序的执行结果如图 11.4 和图 11.5 所示。

图 11.4　不输入标题单击“添加”按钮

图 11.5　输入标题单击“添加”按钮

由于我们尚未完成待办组件的制作，因此会显示组件的默认文字。在下一小节中，我们将待办项组件完善后，基本上就可以看到一个较为完整的待办列表了。

11.3.2　待办项组件的开发

在待办项组件中，最重要的就是组件之间的交互。每个组件的重点在于，既需要从父组件取值，也需要向父组件返回值。接下来打开 item.component.html 输入以下代码：

```
    <nz-card style="width:370px;display:inline-block"
[nzActions]="[actionDone,actionEdit,actionDelete]">
      <nz-card-meta [nzTitle]="title" nzDescription="2018-07-05">
      </nz-card-meta>
    </nz-card>
    <ng-template #actionDone>
      <label nz-checkbox [(ngModel)]="done"
(ngModelChange)="checkItem()"></label>
    </ng-template>
    <ng-template #actionEdit>
      <button (click)="deleteItem()" nz-button nzType="primary">编辑</button>
    </ng-template>
    <ng-template #actionDelete>
      <button (click)="deleteItem()" nz-button nzType="danger">删除</button>
    </ng-template>
```

【代码解析】我们通过设置了一个 nz-card，标题为待办标题，描述设置为时间。然后设置三个 Action 分别为打钩、编辑和删除。由于在主面板组件中，我们还没有增加时间的设置，所以这里先预留好。在后面小节中，我们会逐步完善它。

如前所述，待办项组件的关键就在于与父组件的交互，需要接收的参数为标题、状态、所在的位置。其中所在的位置是用来确定它在数组的位置，以便于删改。在用户对选项进行了修改或删除后，再将选项返回即可。在 item.component.ts 输入以下代码：

```
import { Component, OnInit, Input, Output, EventEmitter } from '@angular/core';
import { NzModalRef, NzModalService } from 'ng-zorro-antd';

@Component({
  selector: 'app-item',
  templateUrl: './item.component.html',
  styleUrls: ['./item.component.scss']
})
export class ItemComponent implements OnInit {
  @Input()
  index: number; // index
  @Input()
  title: string; // 标题
  @Input()
  done: boolean; // 状态
  @Output()
  checkItemEvent = new EventEmitter<Object>(); // 选项打钩事件
  @Output()
  editItemEvent = new EventEmitter<Object>(); // 编辑选项事件
  @Output()
  deleteItemEvent = new EventEmitter<Object>(); // 删除选项事件

  // 对话框
  confirmModal: NzModalRef;
  constructor(private modal: NzModalService) { }
  ngOnInit() { }

  // 选项打钩
  checkItem(): void {
    const data: Object = {
      index: this.index,
      title: this.title,
      done: this.done
    };
    this.checkItemEvent.emit(data);
  }

  // 编辑选项
  editItem() {
    const data: Object = {
      index: this.index,
      title: this.title,
      done: this.done
    };
    this.editItemEvent.emit(data);
  }

  // 删除选项
```

```
  deleteItem(): void {
    this.confirmModal = this.modal.confirm({
      nzTitle: '提示',
      nzContent: '确定要删除吗？',
      nzOkText: '确定',
      nzCancelText: '取消',
      nzOnOk: () => {
        const data: Object = {
          index: this.index,
          done: this.done
        };
        this.deleteItemEvent.emit(data);
      }
    });
  }
}
```

【代码解析】前面已经介绍过所需要接收与返回的参数，而且已经加了注释，所以下面直接讲解方法。选项打钩实质上就是状态修改，checkbox 控件会自动改变自己的状态，所以只要给 ngModelChange 绑定好对应的方法返回即可。editItem 是编辑事件，返回待办项的数据进行编辑。deleteItem 是删除按钮的单击事件，通过一个提示框来进行交互，若单击确认，则返回状态，并在 dashboard 中删除。

最后需要将 dashboard.component.html 和 dashboard.component.ts 进行调整，以适用于我们刚完成的待办选项组件。

```
  // dashboard.component.html
  <div class="dashboard">
    <h2>ToDoList</h2>
    <input nz-input [(ngModel)]="todoTitle" placeholder="添加 ToDo"
class="todo-input">
    <button nz-button nzType="primary" style="margin-left: 15px"
(click)="addTodo()">添加</button>
    <h2>正在进行 {{doingArray.length}}</h2>
    <div *ngFor="let item of doingArray;let i = index" style="margin-top: 15px">
      <app-item [index]="i" [title]="item.title" [date]="item.date"
[done]="item.done"
        (deleteItemEvent)="deleteItem($event)"
(editItemEvent)="editItem($event)"
(checkItemEvent)="checkItem($event)"></app-item>
    </div>
    <h2>已经完成 {{doneArray.length}}</h2>
    <div *ngFor="let item of doneArray;let i = index" style="margin-top: 15px">
      <app-item [index]="i" [title]="item.title" [date]="item.date"
[done]="item.done"
      (checkItemEvent)="checkItem($event)" (editItemEvent)="editItem($event)"
(deleteItemEvent)="deleteItem($event)" ></app-item>
    </div>
  </div>

  // dashboard.component.ts
  import { Component, OnInit } from '@angular/core';
  import { NzMessageService } from 'ng-zorro-antd';
```

```
@Component({
  selector: 'app-dashboard',
  templateUrl: './dashboard.component.html',
  styleUrls: ['./dashboard.component.scss']
})
export class DashboardComponent implements OnInit {

  // 省略...

  // 打钩
  private checkItem(data: Object) {
    const index: number = data['index'];
    const title: boolean = data['title'];
    const done: boolean = data['done'];
    const newItem = {
      done: done,
      title: title
    };
    if (done) {
      this.doingArray.splice(index, 1);
      this.doneArray.push(newItem);
    } else {
      this.doneArray.splice(index, 1);
      this.doingArray.push(newItem);
    }
    this.dataArray = this.doingArray.concat(this.doneArray);
    localStorage.setItem('todo-list', JSON.stringify(this.dataArray));
  }

  // 编辑项目
  editItem(data: Object) {
  }

  // 删除项目
  private deleteItem(data: Object) {
    const index: number = data['index'];
    const done: boolean = data['done'];
    if (done) {
      this.doneArray.splice(index, 1);
    } else {
      this.doingArray.splice(index, 1);
    }
    this.dataArray = this.doingArray.concat(this.doneArray);
    localStorage.setItem('todo-list', JSON.stringify(this.dataArray));
  }

}
```

【代码解析】在 HTML 文件中，对 app-item 标签增加了传值，并绑定 deleteItemEvent 与 checkItemEvent 以接收返回的参数。编辑项目的 editItem 先空着，在下一节完成弹出式窗口功能后再编写。在 TS 文件中，新增了打钩与删除的具体实现方法，实质上都是对数组与 localStorage 进行修改。最终结果如图 11.6 所示。

图 11.6　完成选项组件的待办列表

开发完成后，读者可以多测试一下，看看删除、打钩功能是否能正常使用。如果都没有问题，就继续进入下一小节。下一小节中会将添加功能放到弹出式窗口中，并添加时间与编辑功能。

11.3.3　弹出式窗口组件的开发

虽然我们的输入项只有名称和时间，但为了顺便复习表单的知识，我们还是使用 Angular 的模板驱动型表单来实现。首先需要把 ReactiveFormsModule 导入到 app.module.ts 中，否则接下来可能会报错。

```
   import { BrowserModule } from '@angular/platform-browser';
   import { NgModule } from '@angular/core';

   import { AppComponent } from './app.component';
   import { NgZorroAntdModule, NZ_I18N, zh_CN } from 'ng-zorro-antd';
   import { FormsModule, ReactiveFormsModule } from '@angular/forms';
   import { HttpClientModule } from '@angular/common/http';
   import { NoopAnimationsModule } from '@angular/platform-browser/animations';
   import { registerLocaleData } from '@angular/common';
   import cn from '@angular/common/locales/zh';
   import { DashboardComponent } from
'./components/dashboard/dashboard.component';
   import { ItemComponent } from './components/item/item.component';
   import { EditModalComponent } from
'./components/edit-modal/edit-modal.component';

   registerLocaleData(cn);
```

```
@NgModule({
  declarations: [
    AppComponent,
    DashboardComponent,
    ItemComponent,
    EditModalComponent
  ],
  imports: [
    BrowserModule,
    NgZorroAntdModule,
    FormsModule,
    ReactiveFormsModule,
    HttpClientModule,
    NoopAnimationsModule
  ],
  providers: [{ provide: NZ_I18N, useValue: zh_CN }],
  bootstrap: [AppComponent]
})
export class AppModule { }
```

在 edit-modal.component.html 输入以下代码绘制我们的弹出式窗口。

```
<nz-modal [(nzVisible)]="isVisible" nzTitle="添加待办" nzOkText="确定"
nzCancelText="取消" (nzOnCancel)="handleCancel()"
  (nzOnOk)="handleOk()">
  <form nz-form [formGroup]="validateForm" (ngSubmit)="submitForm()">
    <nz-form-item>
      <nz-form-label [nzSm]="6" [nzXs]="24" nzFor="title" nzRequired>
        <span>
          标题
          <i nz-icon nz-tooltip nzTitle="标题长度不超过 15 位"
type="question-circle" theme="outline"></i>
        </span>
      </nz-form-label>
      <nz-form-control [nzSm]="14" [nzXs]="24">
        <input nz-input id="title" formControlName="title">
        <nz-form-explain *ngIf="validateForm.get('title').dirty &&
validateForm.get('title').errors">请输入标题!
        </nz-form-explain>
      </nz-form-control>
    </nz-form-item>
    <nz-form-item>
      <nz-form-label [nzSm]="6" [nzXs]="24" nzRequired>
        <span>
          日期
          <i nz-icon nz-tooltip nzTitle="不能选择过去的日期"
type="question-circle" theme="outline"></i>
        </span>
      </nz-form-label>
      <nz-form-control [nzSm]="14" [nzXs]="24">
        <nz-date-picker formControlName="date"></nz-date-picker>
        <nz-form-explain *ngIf="validateForm.get('date').dirty &&
validateForm.get('date').errors">请选择日期!
        </nz-form-explain>
      </nz-form-control>
```

```
    </nz-form-item>
  </form>
</nz-modal>
```

【代码解析】我们的弹出式窗口由 nz-modal 构建，在其内部使用 form 制作表单。表单的类型是一个常见的模板驱动型表单，内容包括不超过 15 位的标题与日期。关于模板驱动的内容如果不熟悉了，可以回到第 7 章复习一下，这里不多赘述。

接下来继续实现 edit-modal.component.ts 的代码。

```
import { Component, OnInit, Input, Output, EventEmitter } from '@angular/core';
import { FormBuilder, FormGroup, Validators } from '@angular/forms';

@Component({
  selector: 'app-edit-modal',
  templateUrl: './edit-modal.component.html',
  styleUrls: ['./edit-modal.component.scss']
})
export class EditModalComponent implements OnInit {

  @Input()
  title: string = '';
  @Input()
  date: string = '';
  @Input()
  done: string = '';
  @Input()
  index: number = 0;
  @Input()
  isVisible = false; // 是否显示模态窗
  @Output()
  isVisibleChange = new EventEmitter(); // dialog 显示状态改变事件
  @Output()
  clickEvent = new EventEmitter<Object>();

  isEdit: boolean = false;
  validateForm: FormGroup;

  constructor(private fb: FormBuilder) {
    this.validateForm = this.fb.group({
      title: [null, [Validators.required, Validators.maxLength(15)]],
      date: [null, [Validators.required]]
    });
  }

  ngOnInit() { }

  ngOnChanges() {
    if (this.title) {
      this.isEdit = true;
      this.validateForm.setValue({
        title: this.title,
        date: this.date,
      });
    } else {
```

```
        this.isEdit = false;
        this.validateForm.setValue({
          title: '',
          date: '',
        });
      }
    }

    handleOk(): void {
      this.submitForm();
    }

    handleCancel(): void {
      this.isVisibleChange.emit(false);
    }

    submitForm(): void {
      let params = {};
      for (const i in this.validateForm.controls) {
        this.validateForm.controls[i].markAsDirty();
        this.validateForm.controls[i].updateValueAndValidity();
        if (!(this.validateForm.controls[i].status == 'VALID') &&
this.validateForm.controls[i].status !== 'DISABLED') {
          return;
        }
        if (this.validateForm.controls[i] &&
this.validateForm.controls[i].value) {
          params[i] = this.validateForm.controls[i].value;
        } else {
          params[i] = '';
        }
      }
      this.setDate('date');
      params['date'] = this.validateForm.get('date').value;
      params['isEdit'] = this.isEdit;
      if (this.isEdit) {
        params['done'] = this.done;
        params['index'] = this.index;
      }
      this.clickEvent.emit(params);
      this.isVisibleChange.emit(false);
    }

    // 设置日期格式
    setDate(dates) {
      const time = new Date(this.validateForm.get(dates).value);
      const datetime = time.getFullYear() + '-' +
this.formatDayAndMonth(time.getMonth() + 1) + '-' +
this.formatDayAndMonth(time.getDate());
      this.validateForm.get(dates).setValue(datetime);
    }

    formatDayAndMonth(val) {
      if (val < 10) {
        val = '0' + val;
```

```
    }
    return val;
  }
}
```

【代码解析】这部分的代码比较长，我们一点点分析。从传入参数来看，isVisible 是用于控制显示的，这个肯定是要传值的。另外四个参数实质上是给编辑使用的，所以创建新的待办时不需要传值。之后在构造函数中初始化 FormGroup，并在每次进入页面时使用 ngOnChanges 刷新里边的内容，否则可能会出现旧数据。在返回值之前，时间数据会通过一些方法处理成方便显示的格式，之后把所有参数一起使用 clickEvent 传值回去。

接下来再把 item.component.ts 进行一些修改。

```
import { Component, OnInit, Input, Output, EventEmitter } from '@angular/core';
import { NzModalRef, NzModalService } from 'ng-zorro-antd';

@Component({
  selector: 'app-item',
  templateUrl: './item.component.html',
  styleUrls: ['./item.component.scss']
})
export class ItemComponent implements OnInit {
  @Input()
  index: number; // index
  @Input()
  title: string; // 标题
  @Input()
  date: boolean; // 日期
  @Input()
  done: boolean; // 状态
  @Output()
  checkItemEvent = new EventEmitter<Object>(); // 选项打钩事件
  @Output()
  editItemEvent = new EventEmitter<Object>(); // 编辑选项事件
  @Output()
  deleteItemEvent = new EventEmitter<Object>(); // 删除选项事件

  // 对话框
  confirmModal: NzModalRef;
  constructor(private modal: NzModalService) { }
  ngOnInit() { }

  // 选项打钩
  checkItem(): void {
    const data: Object = {
      index: this.index,
      title: this.title,
      date: this.date,
      done: this.done
    };
    this.checkItemEvent.emit(data);
  }

  // 编辑选项
```

```
    editItem(): void {
      const data: Object = {
        index: this.index,
        title: this.title,
        date: this.date,
        done: this.done
      };
      this.editItemEvent.emit(data);
    }

    // 删除选项
    deleteItem(): void {
      this.confirmModal = this.modal.confirm({
        nzTitle: '提示',
        nzContent: '确定要删除吗？',
        nzOkText: '确定',
        nzCancelText: '取消',
        nzOnOk: () => {
          const data: Object = {
            index: this.index,
            done: this.done
          };
          this.deleteItemEvent.emit(data);
        }
      });
    }
  }
```

【代码解析】页面的接收传值中增加了日期，并且在编辑时都带上这个参数。

接下来要修改 dashboard 的代码，首先修改 dashboard.component.html 和 dashboard.component.scss 文件。

```
  // dashboard.component.html
  <div class="dashboard">
    <h2>ToDoList</h2>
    <div class="top-div">
      <button nz-button nzType="primary" nzBlock (click)="addTodo()">添加待办
</button>
    </div>
    <h2>正在进行 {{doingArray.length}}</h2>
    <div *ngFor="let item of doingArray;let i = index" style="margin-top: 15px">
      <app-item [index]="i" [title]="item.title" [date]="item.date"
[done]="item.done"
        (deleteItemEvent)="deleteItem($event)"
(editItemEvent)="editItem($event)" (checkItemEvent)="checkItem($event)">
      </app-item>
    </div>
    <h2>已经完成 {{doneArray.length}}</h2>
    <div *ngFor="let item of doneArray;let i = index" style="margin-top: 15px">
      <app-item [index]="i" [title]="item.title" [date]="item.date"
[done]="item.done"
        (checkItemEvent)="checkItem($event)"
(editItemEvent)="editItem($event)" (deleteItemEvent)="deleteItem($event)">
      </app-item>
```

```
    </div>
  </div>
  <app-edit-modal [title]="editTitle" [date]="editDate" [done]="editDone"
[index]="editIndex"
    [(isVisible)]="modalIsVisible" (clickEvent)="addTodoEvent($event)">
  </app-edit-modal>

  // dashboard.component.scss
  .dashboard {
    text-align: center;
    color: #2c3e50;
    margin-top: 60px;
  }

  .top-div {
    width: 370px;
    margin-bottom: 15px;
    display: inline-block;
  }

  .item {
    margin: 0px auto;
  }
```

【代码解析】去掉顶部的输入框，改为一个居中的大按钮。之后在底部把我们的弹出式窗口组件显示出来，使用 modalIsVisible 来控制是否弹出这个窗口。

最后将 dashboard.component.ts 的代码修改一下就大功告成了。

```
  ...

    modalIsVisible: boolean = false;
    editTitle: string = '';
    editDate: string = '';
    editDone: boolean = false;
    editIndex: number = 0;

  ...

    // 添加 Todo
    private addTodo(): void {
      this.editTitle = '';
      this.editDate = '';
      this.editDone = false;
      this.editIndex = 0;
      this.modalIsVisible = true;
    }

    // 添加、编辑事件
    addTodoEvent(data: Object) {
      const item = {
        title: data['title'],
        date: data['date'],
        done: data['done']
      };
```

```
    if (data['done'] == true) {
      this.doneArray[data['index']] = item;
    } else {
      if (data['isEdit'] == true) {
        this.doingArray[data['index']] = item;
      } else {
        this.doingArray.push(item);
      }
    }
    this.dataArray = this.doingArray.concat(this.doneArray);
    localStorage.setItem('todo-list', JSON.stringify(this.dataArray));
  }

  // 打钩
  private checkItem(data: Object) {
    const index: number = data['index'];
    const done: boolean = data['done'];
    const newItem = {
      title: data['title'],
      date: data['date'],
      done: done
    };
    if (done) {
      this.doingArray.splice(index, 1);
      this.doneArray.push(newItem);
    } else {
      this.doneArray.splice(index, 1);
      this.doingArray.push(newItem);
    }
    this.dataArray = this.doingArray.concat(this.doneArray);
    localStorage.setItem('todo-list', JSON.stringify(this.dataArray));
  }

  // 编辑项目
  private editItem(data: Object) {
    this.editTitle = data['title'];
    this.editDate = data['date'];
    this.editDone = data['done'];
    this.editIndex = data['index'];
    this.modalIsVisible = true;
  }

...
}
```

【代码解析】这个页面的代码比较多，我们略去了不需要修改的代码部分。首先这个页面增加了多个参数用来给编辑页面传递参数，因为值会和弹出式窗口组件绑定，所以我们在添加时把那几个参数的值清空，这样就可以区分是否为编辑的待办项。现在，保存代码文件，看看浏览器是否如预期的那样成功运行。如果没有问题的话，就可以看到如图 11.7 和图 11.8 所示的运行结果。

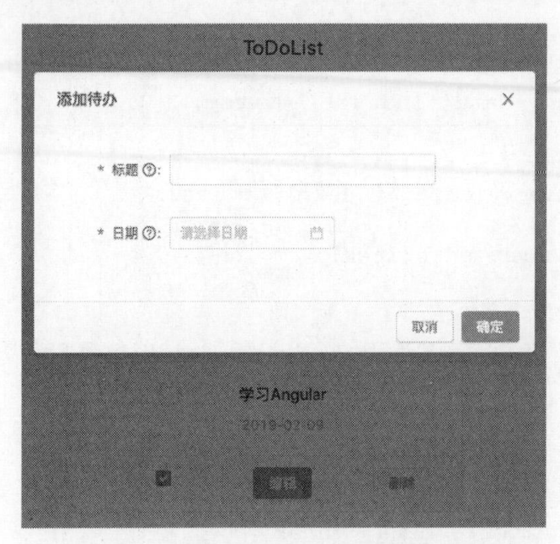

图 11.7　使用弹出式窗口组件添加待办项

图 11.8　使用弹出式窗口组件编辑待办项

11.4　修改为网络请求应用

目前这个项目的实现是基于 LocalStorage 的本地版，接下来将使用 json-server 添加接口并将项目改为带网络请求的应用（网络版）。之所以先用 LocalStorage 实现一次，是因为这个项目完全可以彻底本地化。下面再用 json-server 改写一遍，让我们可以复习一下网络请求相关的知识点。在示例代码下载包中，我们将两份代码分别保存在不同的文件夹中，以方便读者查看。

11.4.1　后台环境的配置

在开发之前，首先要做的就是把后台环境配置好。输入以下代码，并配置 json-server 的 json

文件。

```
json-server --watch data.json

// data.json
{
  "todos": [
    {
      "title": "学习 Angular",
      "date": "2019-02-09",
      "isEdit": true,
      "id": 1,
      "done": false
    },
    {
      "id": 2,
      "title": "学习 TypeScript",
      "date": "2019-02-10",
      "done": true
    }
  ]
}
```

从文件中可以看出，我们设置了 json-server 必须的 id 和待办列表所有必需的参数。接下来只需把对数据的操作改为网络请求即可。

11.4.2　使用 json-server 实现网络请求版

由于文件的数量不是很多，因此这次不制作 HTTP 拦截器，直接调用 HttpClient 发起网络请求。首先修改一下弹出式窗口组件。

```
// edit-modal.component.ts

...
  @Input()
  data: Object = {};
...

  ngOnChanges() {
    if (this.data['id']) {
      this.isEdit = true;
      this.validateForm.setValue({
        title: this.data['title'],
        date: this.data['date'],
      });
    } else {
      this.isEdit = false;
      this.validateForm.setValue({
        title: '',
        date: '',
      });
    }
  }
```

```
...

  submitForm(): void {
    let params = {};
    for (const i in this.validateForm.controls) {
      this.validateForm.controls[i].markAsDirty();
      this.validateForm.controls[i].updateValueAndValidity();
      if (!(this.validateForm.controls[i].status == 'VALID') &&
this.validateForm.controls[i].status !== 'DISABLED') {
        return;
      }
      if (this.validateForm.controls[i] &&
this.validateForm.controls[i].value) {
        params[i] = this.validateForm.controls[i].value;
      } else {
        params[i] = '';
      }
    }
    this.setDate('date');
    params['date'] = this.validateForm.get('date').value;
    params['isEdit'] = this.isEdit;
    if (this.isEdit) {
      params['id'] = this.data['id'];
      params['done'] = this.data['done'];
    } else {
      params['done'] = false;
    }
    this.clickEvent.emit(params);
    this.isVisibleChange.emit(false);
  }
...
```

【代码解析】整体来说代码的改动不大，主要是把传入的多个参数封装为一个 Object 类型的变量，在添加和编辑提交成功时，返回给父组件。没有变化的地方笔者已经标上了省略号，如果有不清楚的地方，可以查阅示例代码。

接下来修改一下待办项组件。

```
// item.component.html
<nz-card style="width: 370px;display: inline-block"
[nzActions]="[actionDone,actionEdit,actionDelete]">
  <nz-card-meta [nzTitle]="data.title" [nzDescription]="data.date">
  </nz-card-meta>
</nz-card>
<ng-template #actionDone>
  <label nz-checkbox [(ngModel)]="data.done"
(ngModelChange)="checkItem()"></label>
</ng-template>
<ng-template #actionEdit>
  <button (click)="editItem()" nz-button nzType="primary">编辑</button>
</ng-template>
<ng-template #actionDelete>
  <button (click)="deleteItem()" nz-button nzType="danger">删除</button>
</ng-template>
```

```
// item.component.ts
...
  @Input()
  data: Object = {};
...
  // 选项打钩
  checkItem(): void {
    this.checkItemEvent.emit(this.data);
  }

  // 编辑选项
  editItem(): void {
    this.editItemEvent.emit(this.data);
  }

  // 删除选项
  deleteItem(): void {
    this.confirmModal = this.modal.confirm({
      nzTitle: '提示',
      nzContent: '确定要删除吗？',
      nzOkText: '确定',
      nzCancelText: '取消',
      nzOnOk: () => {
        this.deleteItemEvent.emit(this.data);
      }
    });
  }
}
```

【代码解析】这个组件同样把传入的多个参数封装为了一个 Object 类型的变量，之后在打钩、编辑、删除时直接将 this.data 传回给父组件即可。由于使用接口进行删改，不需要操作本地数组，因此这个页面不再需要 index 参数来获取自己所在的位置。

最后把最重要的列表组件修改一下，基本上所有的网络请求都是在这个组件中发出的。

接下来修改一下待办项组件。

```
// dashboard.component.html
<div class="dashboard">
  <h2>ToDoList</h2>
  <div class="top-div">
    <button nz-button nzType="primary" nzBlock (click)="addTodo()">添加待办
</button>
  </div>
  <h2>正在进行 {{doingArray.length}}</h2>
  <div *ngFor="let item of doingArray" style="margin-top: 15px">
    <app-item [data]="item" (deleteItemEvent)="deleteItem($event)"
(editItemEvent)="editItem($event)"
      (checkItemEvent)="checkItem($event)">
    </app-item>
  </div>
  <h2>已经完成 {{doneArray.length}}</h2>
  <div *ngFor="let item of doneArray" style="margin-top: 15px">
    <app-item [data]="item" (checkItemEvent)="checkItem($event)"
```

```
(editItemEvent)="editItem($event)"
        (deleteItemEvent)="deleteItem($event)">
      </app-item>
    </div>
  </div>

  <app-edit-modal [data]="editData" [(isVisible)]="modalIsVisible"
(clickEvent)="addTodoEvent($event)">
  </app-edit-modal>

  // dashboard.component.ts
  import { Component, OnInit } from '@angular/core';
  import { NzMessageService } from 'ng-zorro-antd';
  import { HttpClient } from '@angular/common/http';

  @Component({
    selector: 'app-dashboard',
    templateUrl: './dashboard.component.html',
    styleUrls: ['./dashboard.component.scss']
  })
  export class DashboardComponent implements OnInit {

    // 数据数组
    private dataArray: Array<Object> = [];

    // 待办数组
    private doingArray: Array<Object> = [];
    // 完成数组
    private doneArray: Array<Object> = [];

    modalIsVisible: boolean = false;
    editData: Object = {};

    constructor(private message: NzMessageService, private http: HttpClient) { }

    // 生命周期
    ngOnInit() {
      this.getTodoList();
    }

    // 获取列表数据
    private getTodoList(): void {
      this.http.get('http://localhost:3000/todos').subscribe((res:
Array<Object>) => {
        this.dataArray = res;
        this.doingArray = [];
        this.doneArray = [];
        for (let i = 0; i < this.dataArray.length; i++) {
          const element: any = this.dataArray[i];
          if (element.done === true) {
            this.doneArray.push(element);
          } else {
            this.doingArray.push(element);
          }
        }
```

```
      });
    }

    // 添加 Todo
    private addTodo(): void {
      this.editData = {}
      this.modalIsVisible = true;
    }

    // 添加、编辑事件
    addTodoEvent(data: Object) {
      if (data['done'] == true) {
        this.changeItemNetRequest(data);
      } else {
        if (data['isEdit'] == true) {
          this.changeItemNetRequest(data);
        } else {
          this.http.post('http://localhost:3000/todos', data).subscribe((res)
=> {
            this.getTodoList();
          });
        }
      }
    }

    // 打钩
    private checkItem(data: Object) {
      this.changeItemNetRequest(data);
    }

    // 修改的网络请求
    changeItemNetRequest(data: Object) {
      this.http.put('http://localhost:3000/todos/' + data['id'],
data).subscribe((res) => {
        this.getTodoList();
      });
    }

    // 编辑项目
    private editItem(data: Object) {
      this.editData = data;
      this.modalIsVisible = true;
    }

    // 删除项目
    private deleteItem(data: Object) {
      this.http.delete('http://localhost:3000/todos/' +
data['id']).subscribe((res) => {
        this.getTodoList();
      });
    }
  }
```

【代码解析】本页面的代码虽然改动大，但是内容并不复杂。首先导入了 HttpClient 并通过构造函数做了依赖注入。之后在增删改查的方法中，将操作 LocalStorage 改为网络请求即可。对比一下之前的 TS 文件，会发现内容少了很多。最后试一下使用网络请求的这个项目是否能正常运行，由于样式上并没有变化，就不在书中重复提供结果截图了。部分网络请求如图 11.9 所示。

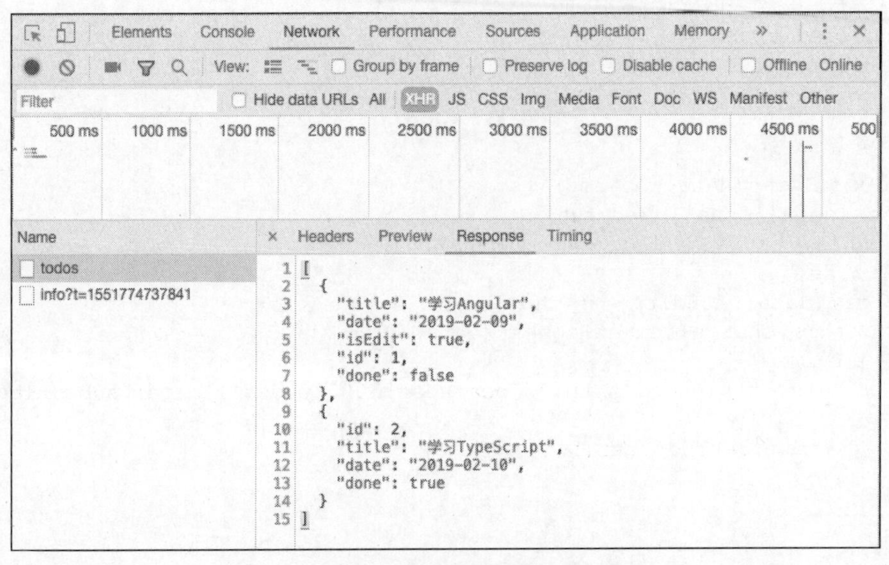

图 11.9 待办列表使用网络请求获取数据

11.5 小 结

本章通过待办列表项目复习了 Angular 的常见知识点，包括 TypeScript 基本语法、组件的使用与传值、内置指令的使用、表单验证等。在完成了本地版的实现后，我们使用 json-server 完成了网络请求版的实现，并复习了网络请求的相关知识。只有通过不断地练习，我们对知识点的掌握才能越来越牢固。下一个实战项目，我们会实现一个功能比较全面的商城后台管理项目，将 Angular 其他相关知识点也用到项目中去。

第12章

项目实战：商城后台管理系统

本章实现一个商城后台管理系统，以应用更多的Angular知识点，并对所学知识点做一个整体复习。我们知道，商城系统一般用于用户在网络上注册登录并购买感兴趣的商品，而商城后台管理系统实质上就是管理员用户对商城的后台商品信息进行维护的工具，比如增加新商品、查看销售额、编辑商品价格等等。因为负责商品信息维护的一般不是开发人员，所以需要为商城后台管理系统开发一套对应的GUI（Graphical User Interface，图形用户界面）以供商品信息维护人员使用。

本章主要涉及的知识点有：

- Angular组件与指令
- Ng-Alain
- G2图标组件
- Angular路由构建
- json-server
- 网络请求

12.1　项目设计

商城类的后台管理系统可以说是屡见不鲜了，市面上已经有很多相对成熟的模板。如果没有特别的想法，按照模板并对照自己的需求设计即可。现在的后台管理系统，基本上都是左边的主导航嵌套子导航，顶部则是搜索框、通知消息、切换主题、退出登录等。这里可以参考一些外国知名的主题模板。Oxygenna Themes网站页面如图12.1所示，EDUMIX网站页面如图12.2所示。

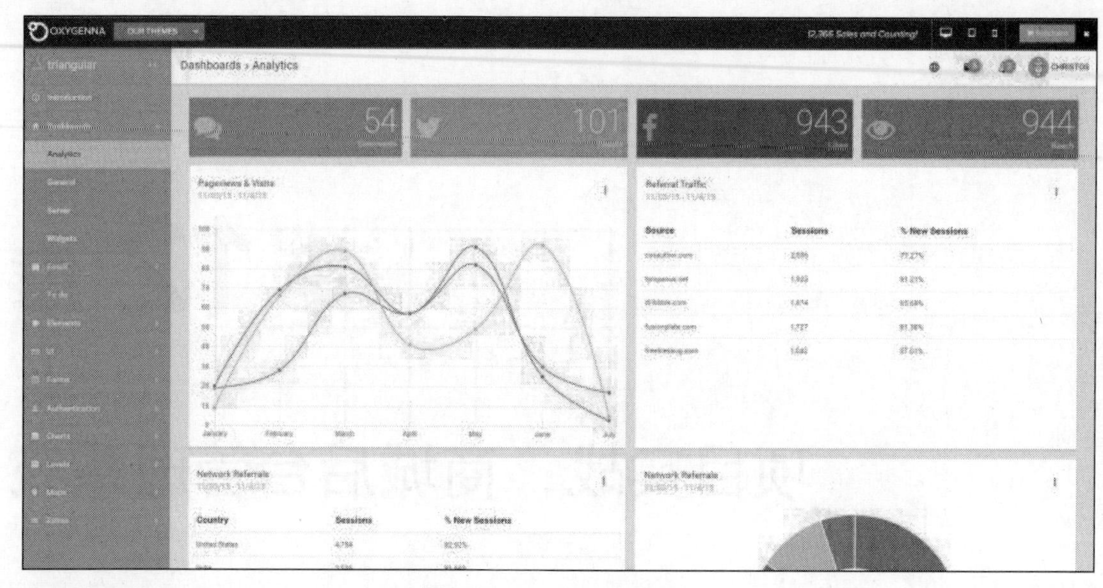

图 12.1　Oxygenna Themes

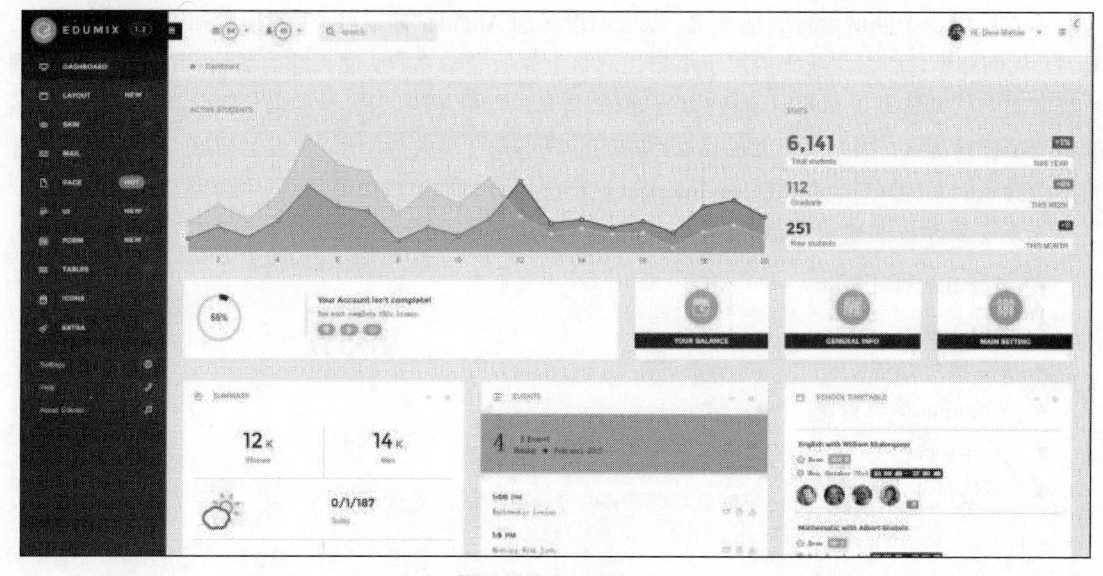

图 12.2　EDUMIX

一个商城后台管理系统，实质上也是增删改查功能。比如新增商品、删除商品、修改商品价格、查询销售记录等。我们的设计同样需要围绕这些实际的业务来展开，一个基本的商城后台管理系统设计基本需要如图 12.3 所示的功能。

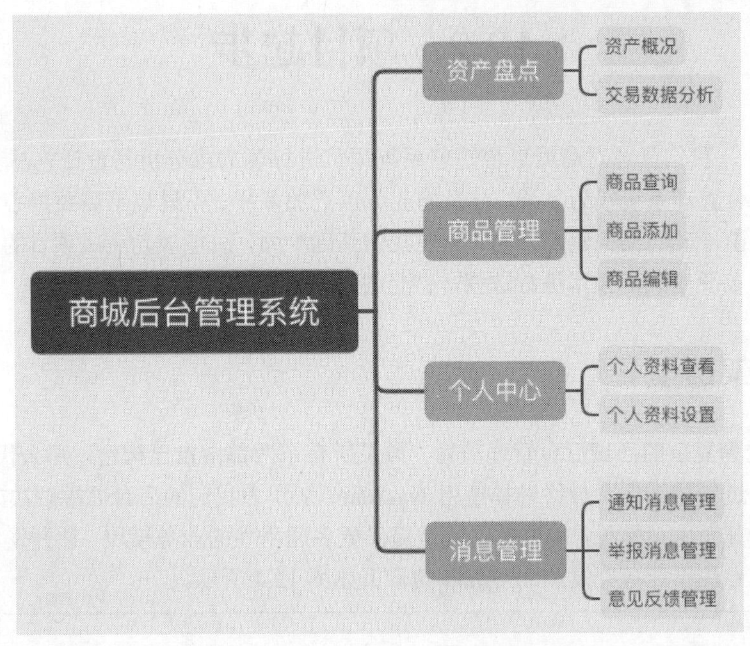

图 12.3　商城后台管理系统设计结构图

我们来梳理一下图 12.3 中的功能分支。

- 资产盘点
  - 资产概况：首页默认显示资产盘点的资产概况，主要展示一些基本概况类的信息。
  - 交易数据分析：销售饼状图、柱形图等。
- 商品管理
  - 商品查询：展示商品列表，通过名称等可以进行查询。
  - 商品添加：新增商品，需要设置商品的名称、价格、图片等。
  - 商品编辑：对已存在的商品进行编辑、修改。
- 个人中心
  - 个人资料查看：查看用户的个人资料等。
  - 个人资料设置：设置用户的名称、头像等资料。
- 消息管理
  - 通知消息管理：查看库存不足等通知、提醒事项。
  - 举报消息管理：查看商品违规等举报消息。
  - 意见反馈管理：查看用户对商品反馈的意见等。

在功能上，这个项目跟真正的商用后台管理软件肯定无法相比，但是作为我们学习 Angular 的练习项目来说，应该是比较全面了。即使这个项目比起商业项目要简单，但是比起上一章的待办列表，它的内容仍然丰富了很多。所以在开发这个项目的时候，我们仍然要全力以赴。

12.2 项目起步

在上一节中，我们已经对商城后台管理系统项目进行了需求分析与设计，与上一章的实战项目不同，商城后台管理系统较为复杂，涉及的业务也更加多样。不过也不需要担心，不管多复杂庞大的项目，也是由一砖一瓦构建起来的，只要我们基础牢固，总能成功完成项目的。在本节中，我们同样先对项目需要用到的技术进行选型，并且进行目录结构、资源的搭建。

12.2.1 框架选型

实现这个较为复杂的商城后台管理项目，如果所有东西都由自己构建，那么代码量会非常大，所以这次笔者经过精挑细选，最终选择使用 Ng-Alain 来作为我们的后台前端解决方案。这种通用解决方案在前端开发中很常见，主要作用就是提供更多通用性的业务模块，让开发者免于重复“发明轮子”，从而专注于业务开发。Ng-Alain 的首页如图 12.4 所示。

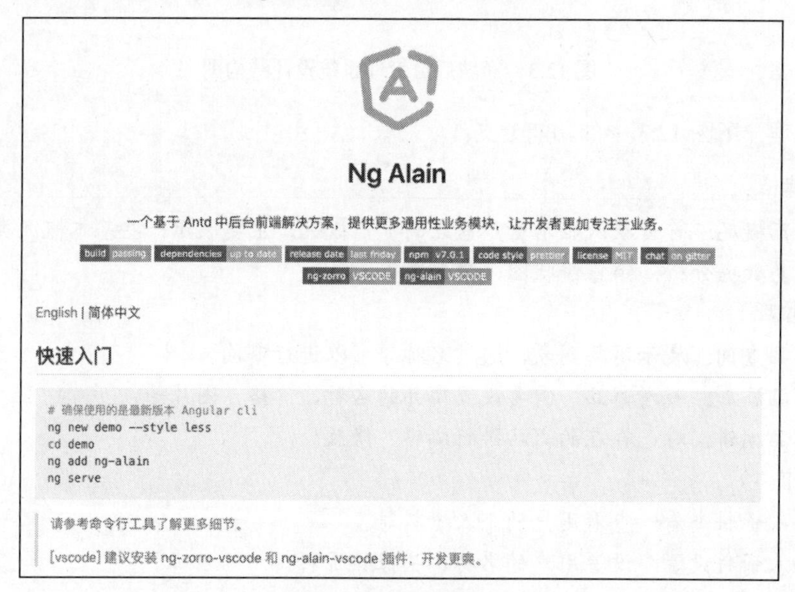

图 12.4 Ng-Alain 首页

值得注意的是，在进行框架选型的时候尽量选择遵循 MIT 协议这种较为宽松的类型，MIT 协议的项目，使用人有权使用、复制、修改、合并、出版发行、散布、再授权、销售软件及其软件的副本。如果是禁止随便使用的框架，请一定要获得授权。

这次还是先对照列举一下 CLI 的版本，便于项目运行出错时进行对比。

```
Angular CLI: 7.1.0
Node: 10.11.0
NPM: 6.4.1
@angular-devkit/architect    0.11.0
@angular-devkit/core         7.1.0
@angular-devkit/schematics   7.1.0
@schematics/angular          7.1.0
```

```
@schematics/update          0.11.0
rxjs                        6.3.3
typescript                  3.1.6
```

笔者的 CLI 版本如上，如果读者在项目运行的过程中出错了，可能是与笔者的版本不一致造成的。

在 UI 样式库中，依然和前面的章节一样，选择使用 NG-ZORRO 进行开发，这次加入了 Ng-Alain，下面也给出版本号。

```
ng-zorro-antd   7.0.0
ng-alain        7.0.1
ng-alain-codelyzer 0.0.1
```

12.2.2　项目的创建

选择好了所用的技术框架，接下来就准备创建项目，输入以下指令创建 TodoList 项目。注意这次使用 CLI 新建项目时，必须选择 LESS，而不是 CSS，因为 Ng-Alain 需要用到。

```
ng new shop
cd shop
ng add ng-alain
```

在使用 Ng-Alain 的过程中，需要对项目做一些设置，我们这里所有的选项都选择为默认选项，如图 12.5 所示。

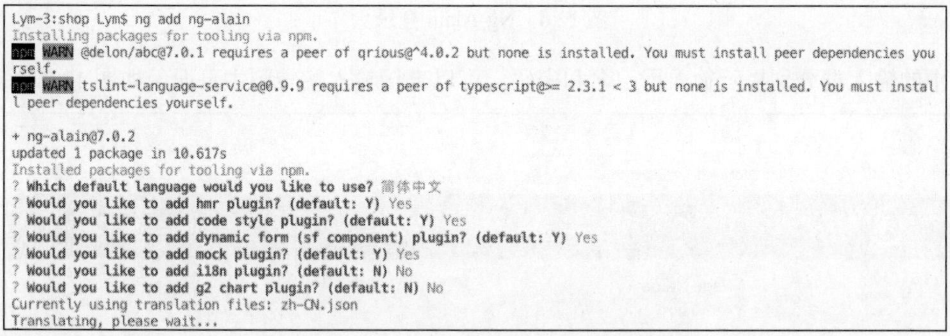

```
Lym-3:shop Lym$ ng add ng-alain
Installing packages for tooling via npm.
npm WARN @delon/abc@7.0.1 requires a peer of qrious@^4.0.2 but none is installed. You must install peer dependencies yourself.
npm WARN tslint-language-service@0.9.9 requires a peer of typescript@>= 2.3.1 < 3 but none is installed. You must install peer dependencies yourself.

+ ng-alain@7.0.2
updated 1 package in 10.617s
Installed packages for tooling via npm.
? Which default language would you like to use? 简体中文
? Would you like to add hmr plugin? (default: Y) Yes
? Would you like to add code style plugin? (default: Y) Yes
? Would you like to add dynamic form (sf component) plugin? (default: Y) Yes
? Would you like to add mock plugin? (default: Y) Yes
? Would you like to add i18n plugin? (default: N) No
? Would you like to add g2 chart plugin? (default: N) No
Currently using translation files: zh-CN.json
Translating, please wait...
```

图 12.5　Ng-Alain 新建项目的设置选择

如果需要使用国际化文字，就需要在"Would you like to add i18n plugin? (default: N)"这个选项中选择 YES。生成的国际化文字文件夹在 assets/tmp/i18n，这个文件夹下存放不同语言文字的配置文件，以中文的配置文件 zh-CN.json 为举例，如下所示。

```
...
  "validation.title.required": "请输入标题",
  "validation.date.required": "请选择起止日期",
  "validation.goal.required": "请输入目标描述",
  "validation.standard.required": "请输入衡量标准"
}
```

【代码解析】这个文件实际上是一个 json 文件，每一列冒号左边的是 Key，右边的则是对应的具体文字。Key 是固定的，在不同语言的配置文件中针对同样的 Key 编写不同的翻译文字，就可以完成了国际化。如果读者需要使用国际化文字，通过以下写法即可完成对 Key 的转换。

```
{{ 'Please enter mobile number!' | translate }}
```

由于大多数国内项目都是固定使用简体中文，所以还是选择默认的不生成配置文件的选项。接下来运行一下项目，运行结果如图 12.6 所示。

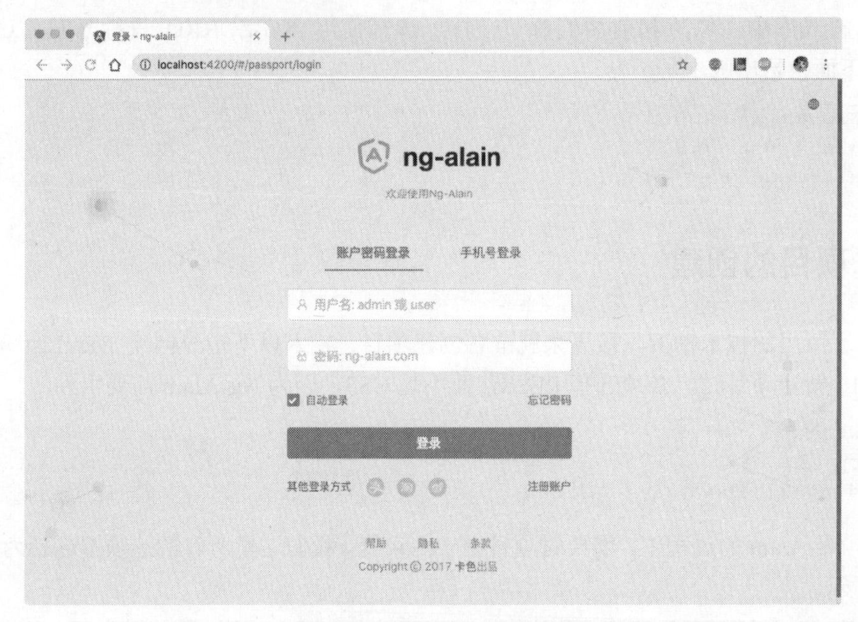

图 12.6　Ng-Alain 登录页

图中的输入框分别提示输入用户名和密码，可以直接输入跳转到主页面，如图 12.7 所示。

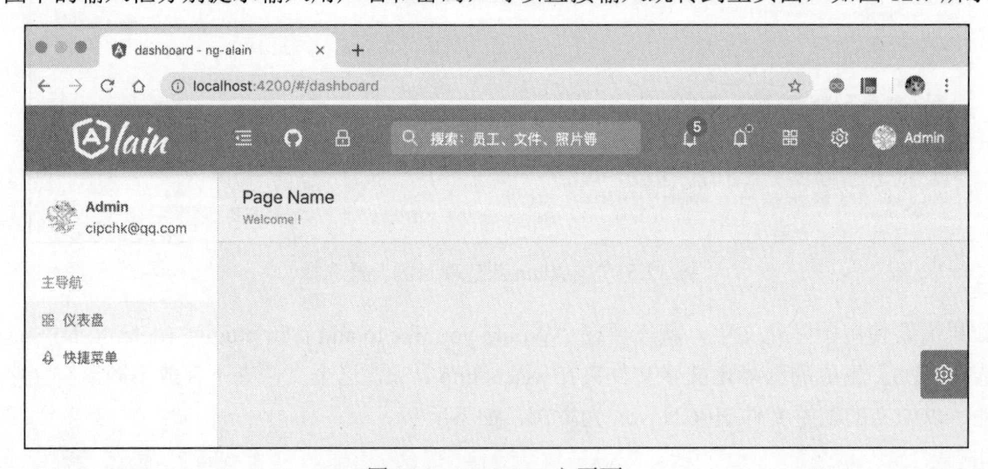

图 12.7　Ng-Alain 主页面

主界面给我们展示的默认样式，基本上与图 12.1、图 12.2 所示的类型相同。目前我们开发的后台管理系统项目，就是建立在这样的一个架构上。之后这个项目的四个大模块逐一完成后，都会添加到左边的主导航中。下一节，我们将正式开始各个模块的开发。

12.3　路由的构建

现在这个项目的主导航部分只有几个空的页面，首先需要把如图 12.3 所示的组件都构建完毕，并按照层级顺序建立自己的路由。至于导航栏上的内容，我们就不动它了，因为这些东西不在我们这次实战的范围内。如果读者想修改的话，可以选择将上面的内容替换为自己想要的内容。

12.3.1　组件的创建

在配置路由之前，把需要用到的组件创建完毕，在终端输入以下指令创建组件。dashboard 组件在项目中已经自带了，所以不需要再重复创建。

```
// dashboard.component.html
ng g component routes/dashboard/dashboard-general
ng g component routes/dashboard/dashboard-echarts

ng g component routes/commodity
ng g component routes/commodity/commodity-search
ng g component routes/commodity/commodity-edit

ng g component routes/person
ng g component routes/person/person-detail
ng g component routes/person/person-setting
```

通过设置路径，将不同模块的组件分别放在了四个不同的父文件夹内，其中商品添加与编辑共用一个组件，而消息管理则复用 NG-Alain 提供的消息组件。组件创建完成后，CLI 会自动在 Module 注册，不需要我们手动注册。最终项目的 src/routes 目录结构如图 12.8 所示。

```
◢ routes
  ▷ callback
  ▷ commodity
  ▷ dashboard
  ▷ exception
  ▷ passport
  ▷ person
  TS routes-routing.module.ts
  TS routes.module.ts
```

图 12.8　routes 的目录结构

12.3.2　路由的配置

平时我们对路由的配置直接在 routes-routing.module.ts 文件中就能完成，但是在 Ng-Alain 中，还需要对 app/core/startup 下的 startup.service.ts 进行配置。这样配置的原因，自然不是为了复杂而复杂。这种构建方式的好处之一就是可以在自己的 service 中动态控制权限，比如切换到非管理员账号，则减少几个选项等。在 routing.module 中则是把所有的东西都注册上，需要哪个就用 service 配置显示哪个。

首先修改 startup.service.ts 中的代码。

```
import { Injectable, Injector, Inject } from '@angular/core';
import { Router } from '@angular/router';
import { HttpClient } from '@angular/common/http';
import { zip } from 'rxjs';
import { catchError } from 'rxjs/operators';
import { MenuService, SettingsService, TitleService, ALAIN_I18N_TOKEN } from
'@delon/theme';
import { DA_SERVICE_TOKEN, ITokenService } from '@delon/auth';
import { ACLService } from '@delon/acl';

import { NzIconService } from 'ng-zorro-antd';
import { ICONS_AUTO } from '../../../style-icons-auto';
import { ICONS } from '../../../style-icons';

/**
 * 用于应用启动时
 * 一般用来获取应用所需要的基础数据等
 */
@Injectable()
export class StartupService {
  constructor(
    iconSrv: NzIconService,
    private menuService: MenuService,
    private settingService: SettingsService,
    private aclService: ACLService,
    private titleService: TitleService,
    @Inject(DA_SERVICE_TOKEN) private tokenService: ITokenService,
    private httpClient: HttpClient,
    private injector: Injector
  ) {
    iconSrv.addIcon(...ICONS_AUTO, ...ICONS);
  }

  private viaHttp(resolve: any, reject: any) {
    zip(
      this.httpClient.get('assets/tmp/app-data.json')
    ).pipe(
      // 接收其他拦截器后产生的异常消息
      catchError(([appData]) => {
        resolve(null);
        return [appData];
      })
    ).subscribe(([appData]) => {

      // application data
      const res: any = appData;
      // 应用信息：包括站点名、描述、年份
      this.settingService.setApp(res.app);
```

```
      // 用户信息：包括姓名、头像、邮箱地址
      this.settingService.setUser(res.user);
      // ACL：设置权限为全量
      this.aclService.setFull(true);
      // 初始化菜单
      this.menuService.add(res.menu);
      // 设置页面标题的后缀
      this.titleService.suffix = res.app.name;
    },
     () => { },
     () => {
       resolve(null);
     });
  }

  private viaMock(resolve: any, reject: any) {
    const app: any = {
      name: `ng-alain`,
      description: `Ng-zorro admin panel front-end framework`
    };
    const user: any = {
      name: 'Admin',
      avatar: './assets/tmp/img/avatar.jpg',
      email: 'cipchk@qq.com',
      token: '123456789'
    };
    // 应用信息：包括站点名、描述、年份
    this.settingService.setApp(app);
    // 用户信息：包括姓名、头像、邮箱地址
    this.settingService.setUser(user);
    // ACL：设置权限为全量
    this.aclService.setFull(true);
    // 初始化菜单
    this.menuService.add([
      {
        text: '主导航',
        group: true,
        children: [
          {
            text: '资产盘点',
            link: '/dashboard',
            icon: { type: 'icon', value: 'anticon-dashboard' },
            children: [
              {
                text: '资产概况',
                link: '/dashboard/general',
              },
              {
                text: '交易数据分析',
                link: '/dashboard/echarts',
              }
            ]
          },
          {
```

```
          text: '商品管理',
          link: '/commodity',
          icon: { type: 'icon', value: 'appstore' },
          children: [
            {
              text: '商品查询',
              link: '/commodity/search',
            },
            {
              text: '商品添加/编辑',
              link: '/commodity/edit',
            }
          ]
        },
        {
          text: '个人中心',
          link: '/person',
          icon: { type: 'icon', value: 'anticon-user' },
          children: [
            {
              text: '个人资料查看',
              link: '/person/detail',
            },
            {
              text: '个人资料设置',
              link: '/person/setting',
            }
          ]
        }
      ]
    }
  ]);
  // 设置页面标题的后缀
  this.titleService.suffix = app.name;
  resolve({});
 }
 load(): Promise<any> {
  return new Promise((resolve, reject) => {
    this.viaMock(resolve, reject);
  });
 }
}
```

【代码解析】虽然代码比较长，但是大多数代码段都含有注释，读者可以参照注释读懂这些代码。menuService.add 方法是用来添加配置的，如果需要添加更多的模块，可以直接在这里添加。

之后就可以回到 routes-routing.module.ts 来配置路由了，代码如下：

```
import { NgModule } from '@angular/core';
import { Routes, RouterModule } from '@angular/router';
import { SimpleGuard } from '@delon/auth';
import { environment } from '@env/environment';
// layout
import { LayoutDefaultComponent } from '../layout/default/default.component';
```

```
    import { LayoutFullScreenComponent } from
'../layout/fullscreen/fullscreen.component';
    import { LayoutPassportComponent } from
'../layout/passport/passport.component';
    // passport pages
    import { UserLoginComponent } from './passport/login/login.component';
    import { UserRegisterComponent } from
'./passport/register/register.component';
    import { UserRegisterResultComponent } from
'./passport/register-result/register-result.component';
    // single pages
    import { CallbackComponent } from './callback/callback.component';
    import { UserLockComponent } from './passport/lock/lock.component';
    // 资产盘点
    import { DashboardComponent } from './dashboard/dashboard.component';
    import { DashboardGeneralComponent } from
'./dashboard/dashboard-general/dashboard-general.component';
    import { DashboardEchartsComponent } from
'./dashboard/dashboard-echarts/dashboard-echarts.component';
    // 商品管理
    import { CommodityComponent } from './commodity/commodity.component';
    import { CommoditySearchComponent } from
'./commodity/commodity-search/commodity-search.component';
    import { CommodityEditComponent } from
'./commodity/commodity-edit/commodity-edit.component';
    // 个人中心
    import { PersonComponent } from './person/person.component';
    import { PersonDetailComponent } from
'./person/person-detail/person-detail.component';
    import { PersonSettingComponent } from
'./person/person-setting/person-setting.component';

    const routes: Routes = [
      {
        path: '',
        component: LayoutDefaultComponent,
        canActivate: [SimpleGuard],
        children: [
          { path: '', redirectTo: 'dashboard', pathMatch: 'full' },
          {
            path: 'dashboard',
            component: DashboardComponent,
            children: [
              {
                path: '',
                component: DashboardGeneralComponent
              },
              {
                path: 'general',
                component: DashboardGeneralComponent
              },
              {
                path: 'echarts',
                component: DashboardEchartsComponent
              }
```

```
          ]
        },
        {
          path: 'commodity',
          component: CommodityComponent,
          children: [
            {
              path: '',
              component: CommoditySearchComponent
            },
            {
              path: 'edit',
              component: CommodityEditComponent
            },
            {
              path: 'search',
              component: CommoditySearchComponent
            }
          ]
        },
        {
          path: 'person',
          component: PersonComponent,
          children: [
            {
              path: '',
              component: PersonDetailComponent
            },
            {
              path: 'detail',
              component: PersonDetailComponent
            },
            {
              path: 'setting',
              component: PersonSettingComponent
            }
          ]
        },
        { path: 'exception', loadChildren:
'./exception/exception.module#ExceptionModule' },
      ]
    },
    // passport
    {
      path: 'passport',
      component: LayoutPassportComponent,
      children: [
        { path: 'login', component: UserLoginComponent, data: { title: '登录',
titleI18n: 'pro-login' } },
        { path: 'register', component: UserRegisterComponent, data: { title: '
注册', titleI18n: 'pro-register' } },
        { path: 'register-result', component: UserRegisterResultComponent, data:
{ title: '注册结果', titleI18n: 'pro-register-result' } },
        { path: 'lock', component: UserLockComponent, data: { title: '锁屏',
titleI18n: 'lock' } },
```

```
    ]
  },
  // 单页不包裹 Layout
  { path: 'callback/:type', component: CallbackComponent },
  { path: '**', redirectTo: 'exception/404' },
];

@NgModule({
  imports: [
    RouterModule.forRoot(
      routes, {
        useHash: false,
        scrollPositionRestoration: 'top',
      }
    )],
  exports: [RouterModule],
})
export class RouteRoutingModule { }
```

【代码解析】这里笔者已经把不同模块做了分类，并添加了注释。在 routes 变量中，需要注意的是不要搞错 path 和 children 等参数，如果发现用鼠标单击了但不跳转到相应的界面，可以回来检查一下这两个参数。

由于默认会显示资产盘点模块，因此先给这个模块添加一下代码，让它可以正常显示自己的子路由。剩余几个大模块的根目录代码除了标题外与这一段代码完全相同，这里就不重复列出来了。

```
// dashboard.component.html
<div class="alain-default__content-title">
  <h1>
    资产盘点
  </h1>
</div>
<router-outlet></router-outlet>
```

运行代码，页面显示效果如图 12.9 所示。

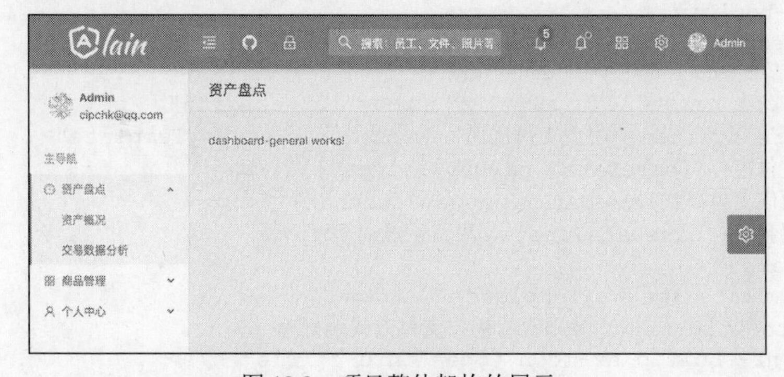

图 12.9　项目整体架构的展示

12.4 资产盘点模块开发

至此路由已经构建完毕，可以继续进行资产盘点模块的开发。资产盘点模块分为资产概况与交易数据分析。在本节中我们会分别完成这两个子功能的开发，包括前端与后端部分。

12.4.1 资产概况开发

资产概况用于展示资产的一些基本概况类的信息，而且该模块会在启动时默认展示。资产概况的文件是 dashboard-general.component.html，首先编写这个文件。

```
    <div nz-row [nzGutter]="24" class="pt-lg">
      <div nz-col nzXs="24" nzSm="12" nzMd="12" nzLg="6">
        <nz-card nzTitle="季度销售额" [nzExtra]="saleTemplate">
          <p *ngFor="let item of baseData.saleroom">{{item.key}}:
{{item.value}}¥</p>
        </nz-card>
        <ng-template #saleTemplate>
          <nz-tooltip nzTitle="查看季度销售额">
            <i nz-tooltip nz-icon type="info-circle"></i>
          </nz-tooltip>
        </ng-template>
      </div>
      <div nz-col nzXs="24" nzSm="12" nzMd="12" nzLg="6">
        <nz-card nzTitle="季度销量" [nzExtra]="volumeTemplate">
          <p *ngFor="let item of baseData.salesVolume">{{item.key}}:
{{item.value}}</p>
        </nz-card>
        <ng-template #volumeTemplate>
          <nz-tooltip nzTitle="查看季度销量">
            <i nz-tooltip nz-icon type="info-circle"></i>
          </nz-tooltip>
        </ng-template>
      </div>
      <div nz-col nzXs="24" nzSm="12" nzMd="12" nzLg="6">
        <nz-card nzTitle="用户支付比例" [nzExtra]="payWayTemplate">
          <p>支付宝：{{baseData?.payway?.alipay}}%</p>
          <p>微信支付：{{baseData?.payway?.wechat}}%</p>
          <p>现金：{{baseData?.payway?.cash}}%</p>
        </nz-card>
        <ng-template #payWayTemplate>
          <nz-tooltip nzTitle="查看用户支付方式偏好">
            <i nz-tooltip nz-icon type="info-circle"></i>
          </nz-tooltip>
        </ng-template>
      </div>
      <div nz-col nzXs="24" nzSm="12" nzMd="12" nzLg="6">
        <nz-card nzTitle="广告转化率" [nzExtra]="adTemplate">
          <p *ngFor="let item of baseData.adRate">{{item}}</p>
        </nz-card>
```

```
    <ng-template #adTemplate>
      <nz-tooltip nzTitle="查看广告转化率">
        <i nz-tooltip nz-icon type="info-circle"></i>
      </nz-tooltip>
    </ng-template>
  </div>
</div>
<nz-card [nzBordered]="false">
  <nz-table #myTable nzBordered [nzData]="listData">
    <thead (nzSortChange)="sort($event)" nzSingleSort>
      <tr>
        <th>商品 id</th>
        <th>商品名</th>
        <th nzShowSort nzSortKey="price">售价</th>
        <th nzShowSort nzSortKey="salesVolume">销量</th>
        <th nzShowSort nzSortKey="inventory">库存</th>
        <th nzShowExpand nzWidth="100px">详细说明</th>
      </tr>
    </thead>
    <tbody>
      <ng-template ngFor let-data [ngForOf]="myTable.data">
        <tr>
          <td>{{data.pId}}</td>
          <td>{{data.name}}</td>
          <td>{{data.price}}</td>
          <td>{{data.salesVolume}}</td>
          <td>{{data.inventory}}</td>
          <td nzShowExpand [(nzExpand)]="mapOfExpandData[data.id]"></td>
        </tr>
        <tr [nzExpand]="mapOfExpandData[data.id]">
          <td colspan="6">{{data.description}}</td>
        </tr>
      </ng-template>
    </tbody>
  </nz-table>
</nz-card>
```

【代码解析】在页面顶部设置了四个卡片组件，分别为季度销售额、季度销量、用户支付比例和广告转化率，并设置了提示信息。页面底部则使用一个列表来展示商品销售信息，其中售价、销量、库存还增加了排序功能。

接下来继续编写 dashboard-general.component.ts 文件。

```
import { Component, OnInit } from '@angular/core';
import { _HttpClient } from '@delon/theme';

@Component({
  selector: 'app-dashboard-general',
  templateUrl: './dashboard-general.component.html',
  styles: []
})
export class DashboardGeneralComponent implements OnInit {
  // 顶部数据
  baseData = {};
  // 列表数据
```

```
    listData = [];
    // 详细说明
    mapOfExpandData = {};
    // 排序名
    sortName = null;
    // 排序值
    sortValue = null;

    constructor(private http: _HttpClient) { }

    ngOnInit() {
      this.getBaseData();
      this.getListData();
    }

    // 获取顶部数据
    getBaseData() {
      this.http.get('http://localhost:3000/inventorLists').subscribe((res) =>
{
        this.baseData = res;
      });
    }

    // 获取列表数据
    getListData() {
      this.http.get('http://localhost:3000/commodity').subscribe((res:
Array<Object>) => {
        this.listData = res;
      });
    }

    // 排序
    sort(sort: { key: string, value: string }): void {
      this.sortName = sort.key;
      this.sortValue = sort.value;
      this.search();
    }

    // 搜索
    search(): void {
      const data = [...this.listData];
      if (this.sortName && this.sortValue) {
        // tslint:disable-next-line:max-line-length
        this.listData = data.sort((a, b) => (this.sortValue === 'ascend') ?
(a[this.sortName] > b[this.sortName] ? 1 : -1) : (b[this.sortName] >
a[this.sortName] ? 1 : -1));
        console.log(this.listData);
      } else {
        this.listData = this.listData;
      }
    }
  }
```

【代码解析】我们为这个页面代码中的各种变量提供了足够多的注释，就不展开讲解了。

getBaseData 和 getListData 这两个方法分别用来获取顶部和列表的展示数据。最后的排序功能需要注意一点，多设置了一个 data 参数并不是多此一举，使用 ES6 的“...”操作符是浅拷贝。如果不设置这个中间变量而直接对 listData 进行排序，那么列表不会进行相应的刷新。

这时候执行代码，我们会得到一个如图 12.10 所示的 404 页面，由于 NG-Alain 框架设置了一个拦截器，如果发起网络请求的时候返回了 404、500 等错误，都会直接跳转到错误页面。

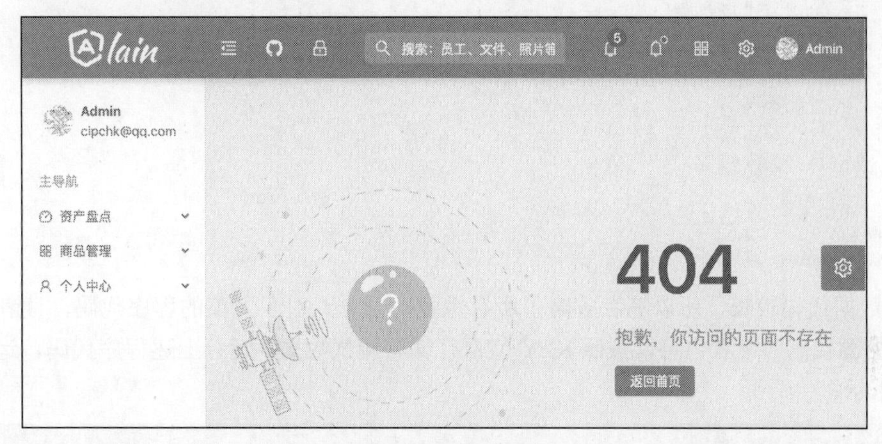

图 12.10　404 页面

最后使用 json-server 设置一下后台数据，页面就能展示出来。使用以下代码启动 json-server，并修改 data.json 文件。

```
json-server --watch data.json

// data.json
{
  "commodity": [
    {
      "id": 1,
      "pId": 1130200088,
      "name": "手机",
      "price": 1500,
      "salesVolume": 2123,
      "inventory": 5200,
      "description": "手机销量良好，请继续保持"
    },
    ...
  ],
  "inventorLists": {
    "saleroom": [
      {
        "key": "一月",
        "value": "32455"
      },
      ...
    ],
    "salesVolume": [
      {
        "key": "一月",
```

```
        "value": "2571"
      },
      ...
    ],
    "payway": {
      "alipay": "43",
      "wechat": "39",
      "cash": "18"
    },
    "adRate": [
      "一月：20%",
      "二月：30%",
      "三月：28%"
    ]
  }
}
```

由于数据代码较长，所以笔者省略了所有重复的部分，对于完整的程序代码，读者可以查看本书配套的源代码。读者也可以按照 KEY 值自行编写测试数据。运行上述程序代码，运行结果如图 12.11 所示。

图 12.11　资产盘点-资产概况页面

12.4.2　交易数据分析的开发

交易数据分析页用于呈现一些销售饼状图、柱形图等。NG-Alain 的图表是在 G2 (3.0) 的基础上二次封装的，提供了业务中常用的图表套件，可以单独使用，也可以组合起来实现复杂的展示效果。如果想要详细学习 G2 图表的一些高级用法，可以参考官方文档 https://antv.alipay.com/zh-cn/g2/3.x/index.html。

首先修改 dashboard-echarts.component.html 文件，代码如下：

```
<nz-card>
  <g2-pie
    [hasLegend]="true"
    title="销售额"
    subTitle="销售额"
    [total]="total"
    [valueFormat]="format"
    [data]="salesPieData"
    height="294">
  </g2-pie>
</nz-card>
<nz-card>
  <g2-bar height="200" [title]="'销售额趋势'" [data]="salesData"></g2-bar>
</nz-card>
```

【代码解析】这个页面代码比较少，主要是展示了一个柱形图和一个饼状图。如果发生报错的情况，请确认安装了 G2 插件，安装指令如下。

```
ng g ng-alain:plugin g2
```

接下来完成它的 TS 文件。

```
import { Component, OnInit } from '@angular/core';
import { _HttpClient } from '@delon/theme';

@Component({
  selector: 'app-dashboard-echarts',
  templateUrl: './dashboard-echarts.component.html',
  styles: []
})
export class DashboardEchartsComponent implements OnInit {

  // 柱状图数据
  salesData = [];
  // 饼状图总额
  total: number = 0;
  // 饼状图数据
  salesPieData = [];
  constructor(private http: _HttpClient) { }
  ngOnInit(): void {
    this.getChartsData();
  }

  // 获取数据
  getChartsData() {
    this.salesData = [];
    this.salesPieData = [];
    this.total = 0;
    this.http.get('http://localhost:3000/echarts').subscribe((res) => {
      const data = res;
      let array = [];
      for (let i = 0; i < 12; i++) {
        const value = data['bar'][i];
```

```
        array.push({ x: `${i + 1}月`, y: value });
      }
      this.salesData = array;
      for (let j = 0; j < data['pie'].length; j++) {
        const obj = data['pie'][j];
        this.salesPieData.push({ x: obj['x'], y: obj['y'], });
        this.total += obj['y'];
      }
    });
  }
}
```

【代码解析】对于这两个图表组件，我们一般需要给它传入一个完整的数组，而不像其他页面通过 ngFor 指令来进行显示。所以在这个页面发出了网络请求后，需要通过循环，将结果数据插入到数组中，最后再刷新页面。

最后在 data.json 文件的尾部插入以下数据，不用管该文件上面的内容。

```
  "echarts": {
    "bar": [
      1000,
      1200,
      1033,
      1222,
      1344,
      2030,
      2800,
      4373,
      3843,
      3933,
      2811,
      2777
    ],
    "pie": [
      {
        "x": "手机",
        "y": 2123
      },
      {
        "x": "电脑",
        "y": 526
      },
      {
        "x": "电视",
        "y": 877
      },
      {
        "x": "空调",
        "y": 222
      },
      {
        "x": "其他",
        "y": 523
      }
    ]
```

```
}
```

运行这段程序代码，结果如图 12.12 所示。

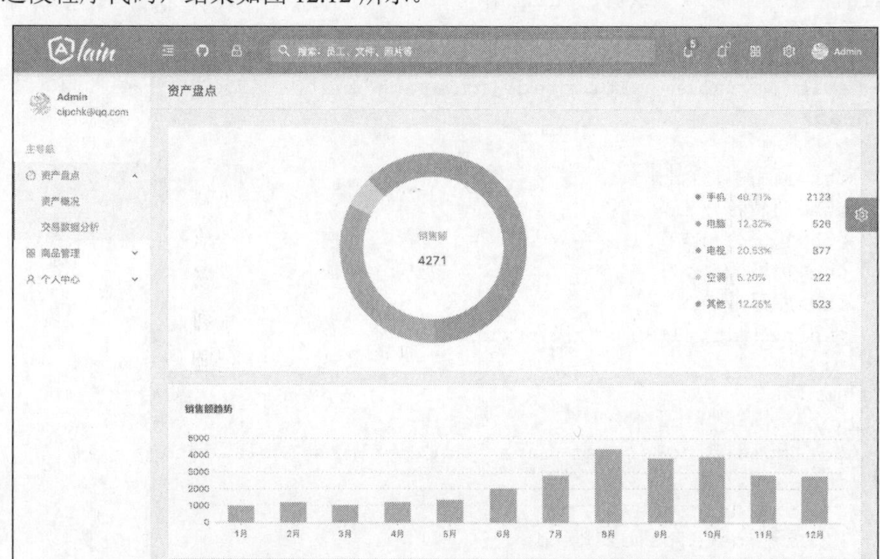

图 12.12　资产盘点-交易数据分析页面

到这里资产盘点模块的开发就完成了，下一节要完成商品管理模块的开发。

12.5　商品管理模块的开发

商品管理模块主要用于对商品信息进行增删改查等相关的操作。此类业务在大多数后台管理系统都会遇到，其功能大同小异，基本上都是列表展示，可以编辑、删除，可以跳转、新增等。商品模块的数据接口与资产概况的接口有一部分是完全相同的，因为对于后台来说，它提供的数据是符合这两个页面的业务要求的，所以可以直接复用。

12.5.1　商品查询的开发

商品查询页除了展示商品列表与删除功能外，还应该提供一个搜索按钮和编辑功能，编辑功能会跳转到添加/编辑页。首先来编写 commodity-search.component.html，代码如下：

```
<nz-card [nzBordered]="false">
  <div>
    <nz-input-group [nzSuffix]="suffixIconSearch">
      <input type="text" nz-input [(ngModel)]="title" placeholder="请输入商品
名查询">
    </nz-input-group>
    <ng-template #suffixIconSearch>
      <i nz-icon type="search"></i>
    </ng-template>
    <button nz-button nzType="primary" (click)="search()"
```

```
style="margin-left:15px">搜索</button>
        <button nz-button nzType="primary" (click)="reset()"
style="margin-left:15px">重置</button>
      </div>
      <br>
      <nz-table #myTable nzBordered [nzData]="listData">
        <thead>
          <tr>
            <th>商品 id</th>
            <th>商品名</th>
            <th>售价</th>
            <th>销量</th>
            <th>库存</th>
            <th nzWidth="180px">操作</th>
          </tr>
        </thead>
        <tbody>
          <ng-template ngFor let-data [ngForOf]="myTable.data">
            <tr>
              <td>{{data.pId}}</td>
              <td>{{data.name}}</td>
              <td>{{data.price}}</td>
              <td>{{data.salesVolume}}</td>
              <td>{{data.inventory}}</td>
              <td>
                <button nz-button nzType="primary" routerLink="/commodity/edit"
[queryParams]="{id:data.id}">编辑</button>
                <button nz-button nzType="danger" (click)="delete(data)">删除
</button>
              </td>
            </tr>
          </ng-template>
        </tbody>
      </nz-table>
    </nz-card>
```

【代码解析】顶部放置了搜索与重置按钮，中间则是一个用于数据展示的表格，它支持编辑、删除等操作，其中编辑功能直接使用了第 8 章讲到的在 HTML 代码中实现路由跳转并传值的操作。

接下来将 TS 文件的代码补充完整。

```
    import { Component, OnInit } from '@angular/core';
    import { _HttpClient } from '@delon/theme';
    import { NzModalRef, NzModalService, NzMessageService } from 'ng-zorro-antd';

    @Component({
      selector: 'app-commodity-search',
      templateUrl: './commodity-search.component.html',
      styles: []
    })
    export class CommoditySearchComponent implements OnInit {
      // 列表数据
      listData = [];
      title = '';
      // 对话框
```

```
    confirmModal: NzModalRef;

    constructor(private http: _HttpClient,
      private message: NzMessageService,
      private modal: NzModalService) { }

    ngOnInit() {
      this.getListData();
    }

    // 获取列表数据
    getListData() {
      let url;
      if (this.title == '') {
        url = 'http://localhost:3000/commodity';
      } else {
        url = 'http://localhost:3000/commodity?name=' + this.title;
      }
      this.http.get(url).subscribe((res: Array<Object>) => {
        this.listData = res;
      });
    }

    // 搜索
    search() {
      this.getListData();
    }

    // 重置
    reset() {
      this.title = '';
      this.getListData();
    }

    // 删除
    delete(data) {
      this.confirmModal = this.modal.confirm({
        nzTitle: '提示',
        nzContent: '确定要删除吗？',
        nzOkText: '确定',
        nzCancelText: '取消',
        nzOnOk: () => {
          this.http.delete('http://localhost:3000/commodity/' +
data['id']).subscribe((res) => {
            this.message.success('删除成功');
            this.reset();
          });
        }
      });
    }
  }
```

【代码解析】在这个页面中使用 title 作为搜索的参数，在使用 getListData 方法获取列表数据时，通过判断 title 是否有值来判断是获取全部数据还只是搜索。后面的删除方法比较简单，弹出式窗口提示框出现后，删除调用接口，最后刷新页面即可。

商品查询部分的内容基本就是这些了，开发完成后读者可以尝试一下搜索和删除功能是否可以正常使用。程序运行的效果如图 12.13 所示。

图 12.13　商品管理-商品查询页面

12.5.2　商品新增/编辑的开发

查询页用于展示和删除商品信息，商品新增/编辑页面主要就是新增和编辑商品信息了。由于新增和编辑同样是操作一个表单，所以我们在一个页面上同时完成两个功能。它们之间的区别则通过 id 来区分。

首先编写 commodity-edit.component.html 的代码。

```
<form nz-form [formGroup]="validateForm" (ngSubmit)="submitForm()">
  <nz-form-item>
    <nz-form-label [nzSm]="6" [nzXs]="24" nzFor="pId" nzRequired>
      <span>
        商品 ID
      </span>
    </nz-form-label>
    <nz-form-control [nzSm]="14" [nzXs]="24">
      <input nz-input id="pId" formControlName="pId">
      <nz-form-explain *ngIf="validateForm.get('pId').dirty &&
validateForm.get('pId').errors">请输入商品 ID!
      </nz-form-explain>
    </nz-form-control>
  </nz-form-item>
  <nz-form-item>
    <nz-form-label [nzSm]="6" [nzXs]="24" nzFor="name" nzRequired>
      <span>
        商品名
      </span>
    </nz-form-label>
```

```
        <nz-form-control [nzSm]="14" [nzXs]="24">
          <input nz-input id="name" formControlName="name">
          <nz-form-explain *ngIf="validateForm.get('name').dirty &&
validateForm.get('name').errors">请输入商品名!
          </nz-form-explain>
        </nz-form-control>
      </nz-form-item>
      <nz-form-item>
        <nz-form-label [nzSm]="6" [nzXs]="24" nzFor="price" nzRequired>
          <span>
            价格
          </span>
        </nz-form-label>
        <nz-form-control [nzSm]="14" [nzXs]="24">
          <input nz-input id="price" formControlName="price">
          <nz-form-explain *ngIf="validateForm.get('price').dirty &&
validateForm.get('price').errors">请输入价格!
          </nz-form-explain>
        </nz-form-control>
      </nz-form-item>
      <nz-form-item>
        <nz-form-label [nzSm]="6" [nzXs]="24" nzFor="salesVolume" nzRequired>
          <span>
            销量
          </span>
        </nz-form-label>
        <nz-form-control [nzSm]="14" [nzXs]="24">
          <input nz-input id="salesVolume" formControlName="salesVolume">
          <nz-form-explain *ngIf="validateForm.get('salesVolume').dirty &&
validateForm.get('salesVolume').errors">请输入销量!
          </nz-form-explain>
        </nz-form-control>
      </nz-form-item>
      <nz-form-item>
        <nz-form-label [nzSm]="6" [nzXs]="24" nzFor="inventory" nzRequired>
          <span>
            库存
          </span>
        </nz-form-label>
        <nz-form-control [nzSm]="14" [nzXs]="24">
          <input nz-input id="inventory" formControlName="inventory">
          <nz-form-explain *ngIf="validateForm.get('inventory').dirty &&
validateForm.get('inventory').errors">请输入库存!
          </nz-form-explain>
        </nz-form-control>
      </nz-form-item>
      <nz-form-item>
        <nz-form-label [nzSm]="6" [nzXs]="24" nzFor="description">
          <span>
            描述
          </span>
        </nz-form-label>
        <nz-form-control [nzSm]="14" [nzXs]="24">
          <input nz-input id="description" formControlName="description">
```

```
    </nz-form-control>
  </nz-form-item>
  <nz-form-item nz-row style="margin-bottom:8px;">
    <nz-form-control [nzSpan]="14" [nzOffset]="6">
      <button nz-button nzType="primary">保存</button>
    </nz-form-control>
  </nz-form-item>
</form>
```

【代码解析】这个页面主要是大量的响应式表单，用于数据验证。除了“描述”字段以外都是必填项。

继续编写 commodity-edit.component.ts 的代码。

```
import { Component, OnInit } from '@angular/core';
import { FormBuilder, FormGroup, Validators } from '@angular/forms';
import { ActivatedRoute, Router } from '@angular/router';
import { _HttpClient } from '@delon/theme';
import { NzMessageService } from 'ng-zorro-antd';

@Component({
  selector: 'app-commodity-edit',
  templateUrl: './commodity-edit.component.html',
  styles: []
})
export class CommodityEditComponent implements OnInit {
  validateForm: FormGroup;
  id: '';

  constructor(private fb: FormBuilder,
    private http: _HttpClient,
    private router: Router,
    private message: NzMessageService,
    private activatedRoute: ActivatedRoute) {
    this.validateForm = this.fb.group({
      id: [null],
      pId: [null, [Validators.required, Validators.pattern(/^[0-9]*$/)]],
      name: [null, [Validators.required]],
      price: [null, [Validators.required, Validators.pattern(/^[0-9]*$/)]],
      salesVolume: [null, [Validators.required,
Validators.pattern(/^[0-9]*$/)]],
      inventory: [null, [Validators.required]],
      description: [null]
    });
  }

  ngOnInit(): void {
    if (this.activatedRoute.snapshot.queryParams['id']) {
      this.id = this.activatedRoute.snapshot.queryParams['id'];
      let url = 'http://localhost:3000/commodity?id=' + this.id;
      this.http.get(url).subscribe((res: Array<Object>) => {
        this.validateForm.setValue({
          id: res[0]['id'],
          pId: res[0]['pId'],
          name: res[0]['name'],
```

```
          price: res[0]['price'],
          salesVolume: res[0]['salesVolume'],
          inventory: res[0]['inventory'],
          description: res[0]['description'] || ''
        });
      });
    } else {
      this.id = '';
      this.validateForm.setValue({
        id: '',
        pId: '',
        name: '',
        price: '',
        salesVolume: '',
        inventory: '',
        description: ''
      });
    }
  }

  submitForm(): void {
    let params = {};
    for (const i in this.validateForm.controls) {
      this.validateForm.controls[i].markAsDirty();
      this.validateForm.controls[i].updateValueAndValidity();
      if (!(this.validateForm.controls[i].status == 'VALID') &&
this.validateForm.controls[i].status !== 'DISABLED') {
        return;
      }
      if (this.validateForm.controls[i] &&
this.validateForm.controls[i].value) {
        params[i] = this.validateForm.controls[i].value;
      } else {
        params[i] = '';
      }
    }
    if (this.id == '') {
      this.http.post('http://localhost:3000/commodity',
params).subscribe((res) => {
        this.message.success('添加成功');
        this.router.navigate(['/commodity/search']);
      });
    } else {
      this.http.put('http://localhost:3000/commodity/' + this.id,
params).subscribe((res) => {
        this.message.success('编辑成功');
        this.router.navigate(['/commodity/search']);
      });
    }
  }
}
```

【代码解析】本页面通过是否有 id 来判断执行新增还是编辑操作。如果为编辑操作，就先通过 URL 中的 id 发送网络请求，获取商品详情，再给表单附加默认值。在验证项中，商品编号、价格等必须为纯数字，因为在盘点页中涉及排序。最后在完成表单提交操作后，显示操作成功的提示信息并跳转回商品查询页。编辑页的显示如图 12.14 所示。

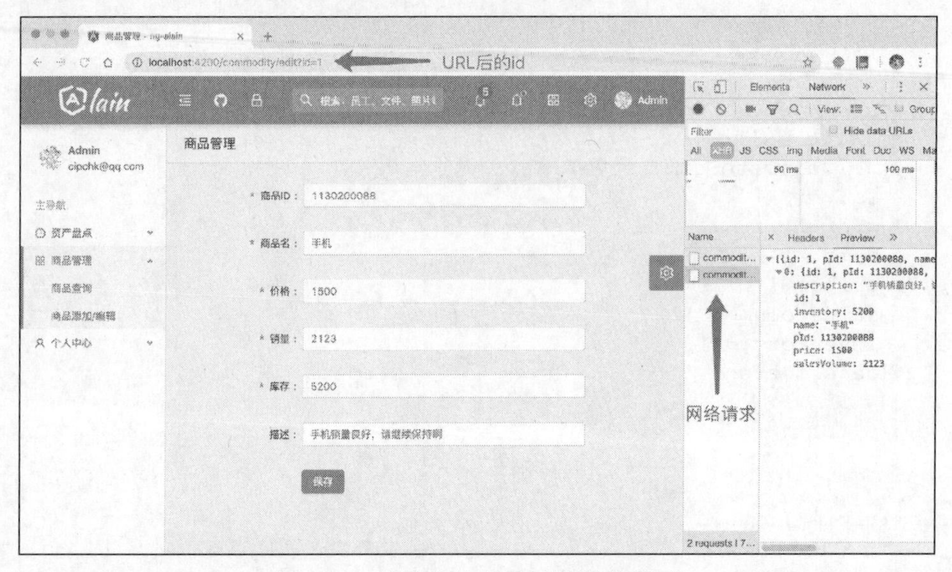

图 12.14　商品管理-商品新增/编辑页面

从图 12.14 中箭头指向可以看到，编辑时 URL 会携带商品 id，并通过网络请求获取商品详情。接下来我们还是先将编辑与添加功能调试成功，成功后，在下一节，我们将继续完成个人中心模块的开发。

12.6　个人中心模块的开发

至此，路由已经构建完毕，我们继续进行资产盘点模块的开发。资产盘点模块分为资产概况与交易数据分析。接下来会分两个小节来完成这部分的开发，包括前端与后端部分。

12.6.1　个人资料查看的开发

任何软件都是给人操作使用的，所以软件基本上也绕不开用户信息的设置与查看。先修改 person-detail.component.html 文件，代码如下：

```
<nz-card [nzBordered]="false" class="mb-lg">
  <div class="avatarHolder">
    <img style="width:100px;height:100px;" [src]="userInfo?.avatar">
    <nz-divider nzDashed></nz-divider>
    <div>姓名：{{userInfo?.username}}</div>
    <div>身份：{{userInfo?.status}}</div>
    <div>个性签名：{{userInfo?.sign}}</div>
  </div>
  <nz-divider nzDashed></nz-divider>
  <div class="tags">
```

```
    <nz-tag [nzColor]="'#f50'">技术控</nz-tag>
    <nz-tag [nzColor]="'#2db7f5'">管理员</nz-tag>
    <nz-tag [nzColor]="'#87d068'">业绩优秀</nz-tag>
    <nz-tag [nzColor]="'#108ee9'">步步高升</nz-tag>
  </div>
</nz-card>
```

【代码解析】内容不是很多，通过双向绑定展示用户数据。

继续编写实现文件的代码，并提前为 data.json 文件添加一些测试数据以方便展示。

```
// person-detail.component.ts
import { Component, OnInit } from '@angular/core';
import { _HttpClient } from '@delon/theme';

@Component({
  selector: 'app-person-detail',
  templateUrl: './person-detail.component.html',
  styles: []
})
export class PersonDetailComponent implements OnInit {

  userInfo: Object = {};
  constructor(private http: _HttpClient) { }
  ngOnInit() {
    this.getUserInfo();
  }

  getUserInfo() {
    this.http.get('http://localhost:3000/user').subscribe((res) => {
      this.userInfo = res;
    });
  }
}

// data.json
...
  "user": {
    "avatar": "https://avatars1.githubusercontent.com/u/16334445?s=460&v=4",
    "username": "张三",
    "status": "管理员",
    "sign": "大家好啊"
  }
```

【代码解析】该页面导入 HttpClient 发送网络请求获取用户数据，通过 userInfo 接收数据并绑定在页面中。程序的执行结果如图 12.15 所示。

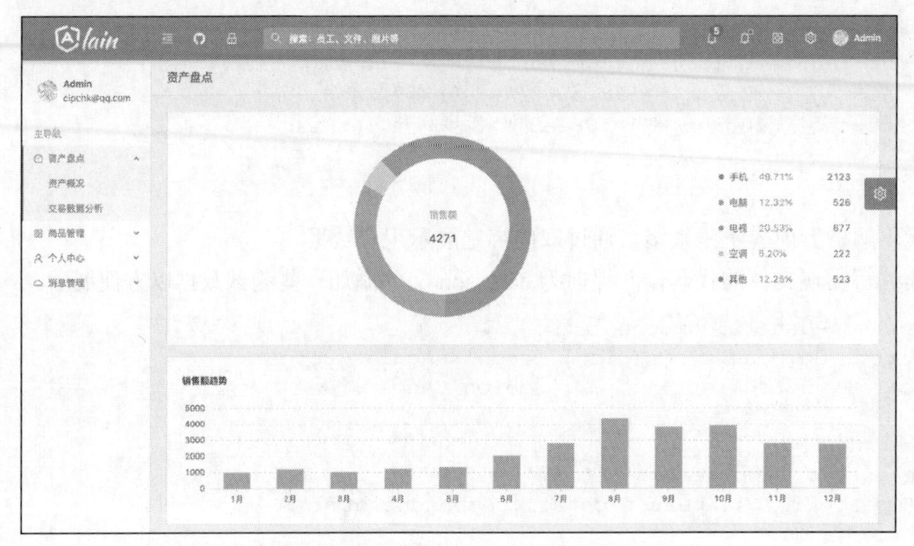

图 12.15　个人中心-个人资料查看页面

12.6.2　个人资料设置的开发

在展示完个人信息后，我们会联想到还需要修改个人信息的功能，否则的话这个系统肯定算不上完善。个人资料设置内容同样不是很复杂，通过一个表单页提交数据即可。资产概况的文件是person-setting.component.html，首先编写这个文件。

```
    <form nz-form [formGroup]="validateForm" (ngSubmit)="submitForm()">
      <nz-form-item>
        <nz-form-label [nzSm]="6" [nzXs]="24" nzFor="avatar" nzRequired>
          <span>
            头像 URL
          </span>
        </nz-form-label>
        <nz-form-control [nzSm]="14" [nzXs]="24">
          <input nz-input id="avatar" formControlName="avatar">
          <nz-form-explain *ngIf="validateForm.get('avatar').dirty &&
validateForm.get('avatar').errors">请输入头像 URL!
          </nz-form-explain>
        </nz-form-control>
      </nz-form-item>
      <nz-form-item>
        <nz-form-label [nzSm]="6" [nzXs]="24" nzFor="username" nzRequired>
          <span>
            姓名
          </span>
        </nz-form-label>
        <nz-form-control [nzSm]="14" [nzXs]="24">
          <input nz-input id="username" formControlName="username">
          <nz-form-explain *ngIf="validateForm.get('username').dirty &&
validateForm.get('username').errors">请输入姓名!
          </nz-form-explain>
        </nz-form-control>
```

```
      </nz-form-item>
      <nz-form-item>
        <nz-form-label [nzSm]="6" [nzXs]="24" nzFor="status" nzRequired>
          <span>
            身份
          </span>
        </nz-form-label>
        <nz-form-control [nzSm]="14" [nzXs]="24">
          <nz-select id="status" formControlName="status" nzPlaceHolder="请选择
">
            <nz-option nzValue="普通用户" nzLabel="普通用户"></nz-option>
            <nz-option nzValue="管理员" nzLabel="管理员"></nz-option>
          </nz-select>
          <nz-form-explain *ngIf="validateForm.get('status').dirty &&
validateForm.get('status').errors">请选择身份!
          </nz-form-explain>
        </nz-form-control>
      </nz-form-item>
      <nz-form-item>
        <nz-form-label [nzSm]="6" [nzXs]="24" nzFor="sign">
          <span>
            个性签名
          </span>
        </nz-form-label>
        <nz-form-control [nzSm]="14" [nzXs]="24">
          <input nz-input id="sign" formControlName="sign">
        </nz-form-control>
      </nz-form-item>
      <nz-form-item nz-row style="margin-bottom:8px;">
        <nz-form-control [nzSpan]="14" [nzOffset]="6">
          <button nz-button nzType="primary">保存</button>
        </nz-form-control>
      </nz-form-item>
    </form>
```

【代码解析】除了控制身份的参数 status 外，其他参数都属于普通的 input 输入项。头像 URL 这个参数除了可以设置网上可用的链接外，也可以通过绝对路径放置本地的图片。

最后把 TS 文件中的代码补全，这个模块也就大功告成了。

```
import { Component, OnInit } from '@angular/core';
import { FormBuilder, FormGroup, Validators } from '@angular/forms';
import { ActivatedRoute, Router } from '@angular/router';
import { _HttpClient } from '@delon/theme';
import { NzMessageService } from 'ng-zorro-antd';

@Component({
  selector: 'app-person-setting',
  templateUrl: './person-setting.component.html',
  styles: []
})
export class PersonSettingComponent implements OnInit {
  validateForm: FormGroup;

  constructor(private fb: FormBuilder,
```

```
    private http: _HttpClient,
    private router: Router,
    private message: NzMessageService,
    private activatedRoute: ActivatedRoute) {
    this.validateForm = this.fb.group({
      avatar: [null, [Validators.required]],
      username: [null, [Validators.required]],
      status: [null, [Validators.required]],
      sign: [null]
    });
  }

  ngOnInit() {
    this.http.get('http://localhost:3000/user').subscribe((res) => {
      this.validateForm.setValue({
        avatar: res['avatar'],
        username: res['username'],
        status: res['status'],
        sign: res['sign'] || ''
      });
    });
  }

  submitForm() {
    let params = {};
    for (const i in this.validateForm.controls) {
      this.validateForm.controls[i].markAsDirty();
      this.validateForm.controls[i].updateValueAndValidity();
      if (!(this.validateForm.controls[i].status == 'VALID') &&
this.validateForm.controls[i].status !== 'DISABLED') {
        return;
      }
      if (this.validateForm.controls[i] &&
this.validateForm.controls[i].value) {
        params[i] = this.validateForm.controls[i].value;
      } else {
        params[i] = '';
      }
    }
    this.http.put('http://localhost:3000/user/', params).subscribe((res) => {
      this.message.success('修改成功');
      this.router.navigate(['/person/detail']);
    });
  }
}
```

【代码解析】本页面在进入时会先调用接口获取默认数据，提交的时候会通过 FormGroup 来校验输入项是否准确、完整。完成后弹出用于显示成功提示信息的窗口，并通过路由跳转回个人资料查看页面。页面显示效果如图 12.16 所示。

总的来说，这个模块内容与商品管理模块十分相似，部分代码都是可以直接复用的，这也就是为什么许多人笑称程序员干的大多数工作其实是重复性的工作。实际上，在开发实际项目中同样如此，在完成核心功能后，大多数时间都是在干使用重复代码类似的工作。即便如此，我们也要尽可能保证代码的质量，不能因为复制粘贴快速完事而导致项目出现大量的 BUG（程序错误）。

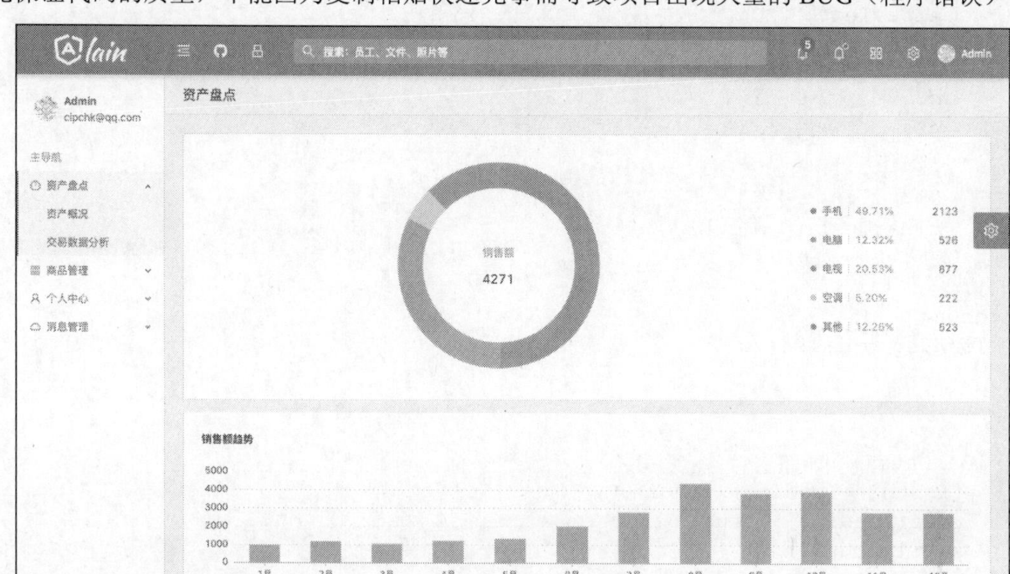

图 12.16　个人中心-个人资料设置页面

12.7　消息管理模块的开发

在后台管理系统应用中，通常都会有一个消息通知模块。这种功能一般用于发布一些通知、待办类消息，以提醒管理员及时处理。在前面的各个页面中我们就可以看到，在顶部的导航栏中，已经配置好了一个待办消息通知的小铃铛图标，我们这个模块会使用 NG-Alain 提供的模板进行二次开发。

在消息管理模块中，我们设置了通知、举报和待办三种消息，实质上并不需要为这三种类型的消息设置三个页面。它们实质上都是消息，只是被标记为不同的类型。这次要修改的文件夹在 src/app/layout/default/header/components/notify.component.ts，如果读者觉得在层级比较多的文件夹中定位文件效率低下，当然可以使用 Ctrl+P 键（Mac 是 Command+P 键）来快速找到文件。

这个文件 NG-Alain 已经为我们配置完毕，我们只要直接对其进行修改即可，这个文件的作用是展示我们设计的三种消息。

```
...
export class HeaderNotifyComponent {
  data: NoticeItem[] = [
    {
      title: '通知',
      list: [],
      emptyText: '无',
      clearText: '清空消息',
    },
    {
      title: '举报',
```

```
      list: [],
      emptyText: '无',
      clearText: '清空消息',
    },
    {
      title: '待办',
      list: [],
      emptyText: '无',
      clearText: '清空消息',
    },
  ];
  count = 5;
  loading = false;

...

  loadData() {
    if (this.loading) return;
    this.loading = true;
    setTimeout(() => {
      this.data = this.updateNoticeData([
        {
          id: '000000001',
          title: '手机货物不足，请及时补货',
          datetime: '2019-03-09',
          type: '通知',
        },
        {
          id: '000000002',
          title: '本周周末加班',
          datetime: '2019-03-09',
          type: '通知',
        },
        {
          id: '000000003',
          title: '张三的举报',
          description: '举报举报举报举报举报举报',
          datetime: '2019-03-09',
          type: '举报',
        },
        {
          id: '000000004',
          title: '李四的举报',
          description: '举报举报举报举报举报举报',
          datetime: '2019-03-09',
          type: '举报',
        },
        {
          id: '000000005',
          title: '全部收收',
          description: '任务需要在 2019-04-09 20:00 前启动',
          extra: '未开始',
          status: 'todo',
          type: '待办',
```

```
        },
        {
          id: '000000006',
          title: '下个月前销售额必须翻倍',
          description:
            '2019-03-01',
          extra: '进行中',
          status: 'urgent',
          type: '待办',
        },
      ]);
      this.loading = false;
      this.cdr.detectChanges();
    }, 1000);
  }

  clear(type: string) {
    this.msg.success(`清空了 ${type}`);
  }

  select(res: any) {
    this.msg.success(`单击了 ${res.title} 的 ${res.item.title}`);
  }
}
```

【代码解析】在这个页面中，我们关键要修改的是两组数据。一个是用来控制通知的类型，我们添加了通知、举报、待办三种。另一个是用来控制显示的消息列表，这个消息全部写在一起，通过 type 来区分类型。修改完毕后，运行结果如图 12.17 所示。

图 12.17　消息管理模块页面

本节的代码较多，笔者省略了一些没有变动的代码，如果需要查看完整的程序代码，可以查阅本书配套的示例代码文件。

12.8 小 结

本章通过实现一个完整的商城后台系统，学习了项目设计、项目构建、路由配置，并在后面的不同模块中运用了组件、路由、网络请求、指令等知识点。本书中的项目例子较多，因为笔者相信只有通过多练多写，才能在学习编程的过程中取得更快的进步。只是看看书、翻翻 API 文档而不动手实践的话，很可能会坚持不下去，学习效果也不好。最后，希望读者通过对本书例子的练习，更加牢固地掌握 Angular 中的知识点，不断提高自己的前端开发技能。